Rat-Free Leadership

RAT FREE
LEADERSHIP

KILL CONFUSION

DRIVE ENGAGEMENT

BUILD AN UNSTOPPABLE TEAM

CURTISS MURPHY

Illustrations and cover by Ivano Nardacchione

Interior Layout by Brittany Becker

RAT-FREE LEADERSHIP
Kill Confusion, Drive Engagement, and Build an Unstoppable Team.

First Edition

ISBN: 979-8-9931486-0-1 Paperback
IBSN: 979-8-9931486-1-8 EBook

Kayla,
For Gratitude

Jacob,
For Joy

Jenny,
For Love

Table of Contents

Chapter 1

Who's To Blame?

C rap!"

I said out loud. To an empty office.

How could this have happened? Our teams were careful. Committed.

Still, we missed it.

I called a big meeting to break the news and figure out next steps.

This was early in COVID. We'd gone fully remote. Slack. Zoom. All of it.

We were working on a project for one of the biggest companies in the world. You'd know them.

Let's call them BigMegaCorp.

They wanted us to develop multiple games—ASAP. They said, "Jump." We did.

Threw out the whole roadmap.

Didn't even have a finalized contract. We disrupted two—no, three—teams.

And wow. They pivoted like champs.

Fluid. Collaborative. Engaged. Moving as one. Staying ahead of schedule.

Things were looking great. And we still had over three weeks left. Fantastic. Almost exceptional.

"Almost," because we missed something.

A *tiny* clause.

The launch date wasn't the date that mattered. BigMegaCorp wanted the builds three weeks *early*.

For their own reviews.

"Crap!"

★ ★ ★

"Thanks for coming. Sorry to interrupt the work. I'll cut to the chase."

I took a breath. All eyeballs locked on me.

"We have a problem."

Then, I explained the hidden fine print. We didn't have three weeks. We had until Friday.

It was Tuesday.

Three days left. Tens of millions on the line.

★ ★ ★

So what do you think? Leader to leader.

Who's at fault here?

Was it me? I ran art, testing, engineering. I knew the teams. The work. Execution.

Was it the ICs? The individual contributors? Obviously not. A woodworker doesn't blame the lathe.

Marketing? They handled the deal. Saw the fine print.

Or product? They chose the features. Set priorities. Owned the roadmap.

Who's to blame?

That's the question a lot of leaders ask in moments like this. Who dropped the ball?

It's the *wrong* question.

Because this wasn't a mediocre team. We had engaged staff. Proven processes. Excellent communication.

No conflicts. No weak links. Teams that did the things I'm about to teach you.

So, asking who's to blame? Nope.

The root cause wasn't a "who." It wasn't a broken process.

It was one awful word.

Confusion.

★ ★ ★

You want something to sink your teeth into? Here's the most important lesson in this book:

Confusion is everywhere.

Again!

Confusion is everywhere. All the time. It can never be eradicated.

That's the honest truth.

Confusion is the real enemy. It destroys performance. It kills teams. It kills people.

Figuratively. And, in healthcare, literally. I'll show you.

For now, let's focus.

We're not talking about call centers or factories.

We're talking about creative work. Fast-paced, flexible, competitive.

We pivot daily. And every time we do, cracks form. Tiny gaps of information.

Knowledge is lost.

In the case of BigMegaCorp? The crack started with one line:

"Didn't even have a finalized contract."

That was the error.

Sure, in hindsight, it's obvious. Maybe you saw it right away. You're sitting there. Reading the breadcrumbs

That's not how life works.

In the real world, there are no breadcrumbs. There's no narrator dropping hints. You can't see the cracks as they form. And they're there.

So, was it preventable?

Of course!

Someone knew the correct date. I'm sure a few people even talked about it. Those people *assumed* everyone knew.

These weren't incompetent people. Or disengaged. Or careless.

It's just what happens with fast work, pivots, and complex systems that are changing. Critical pieces of information *will* fall through the cracks.

Because confusion is everywhere. All the time.

Confusion is when two people "know" different things.

The space between what a person knows and what they *assume* others know.

It lives in change. In pivots. In speed.

Confusion is the enemy of excellence. And if you want to build unstoppable teams, then you need to do everything in your power to destroy it.

Knowing full well you'll never succeed.

You can never beat confusion.

I'll prove it to you.

★ ★ ★

Before I do, here's some bona fides.

I survived a brutal industry.

Three decades of leadership. Thousands of game releases. Three hundred *million* players.

Those years in the trenches taught me important lessons.

I had to ditch the '90s playbook, quit serving crap-sandwiches, and face the elephants hiding in plain sight.

I built a better playbook:

- **Ruthless Clarity**
- **Strong Performers**
- **Relentless Engagement**

That's how great leaders build unstoppable teams.

Now, back to Whisper—the rat destroying everything.

P.S. I have lots of stories. Sometimes I change the details or mash people together. Unless I'm talking about Adonis—he's real. Mostly.

Chapter 2

When Death is on the Line

My obsession with confusion began thirteen years ago.

"Curtiss! Let's go to lunch."

My boss's boss was a retired navy admiral.

Fit, polished, clever. He had that smooth-talking charm. The kind that's standard issue in the "New Admirals, Welcome Aboard" packet.

He came out swinging, before we even got to the restaurant.

"So ... Curtiss. What do you think about healthcare?"

What a weird question.

"Healthcare?" I asked.

That's my best response? Way to go, Sparky McFly.

The admiral chuckled. He looked mischievous.

"Yes. Healthcare."

He paused dramatically. "I have a proposal for you."

★ ★ ★

That's how I ended up in medical school.

One minute I'm a project lead, trying to understand why games work.

Next, I'm carrying a box (keyboard, mouse, cables), walking down the hall of a beautiful building.

The Simulation Center at Eastern Virginia Medical School (EVMS).

And I'm not visiting. I'm finding my office—a swanky setup, right next to the dean.

Outside my door was a fancy plaque.

The top was polished, inlaid wood. A permanent fixture, with a stenciled number.

The bottom part stopped me cold. In big, bold letters was my name.

Somehow, the ex-admiral had formed a partnership with EVMS. Got me VIP access to almost anything on campus.

For an entire year.

Even for him, this was next-level.

I asked, "What am I supposed to do there? I'm not a med student, I'm a programmer."

His answer was cryptic.

"Learn."

No requirements. No deliverables. Unlimited access to medical technology.

★ ★ ★

That's how I made a terrible discovery. A lesson that changed my path as a leader.

A deep, dark truth.

Once you hear it, you can't unhear it.

Now, you might be in healthcare. If you're a doctor, a nurse—especially a nurse, they do the hard work. If you're in healthcare, you've heard this.

If not, I'm sorry. I'm going to share something dark.

It's this:

One in twenty-five people will die from human error.

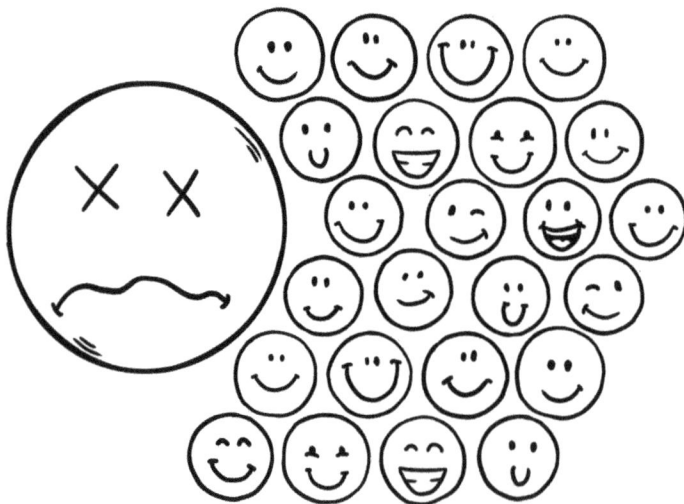

Let's make this real.

For every hundred people reading this sentence, four of them will die from human error.

You. Me. Everyone. One in twenty-five odds.

It gets worse.

These aren't deaths from surgical complications. Or complex diagnoses.

No. These are deaths from basic breakdowns in communication. Failures of teamwork.

One person knew something and didn't speak up. Kind of like the date in the contract.

Someone didn't communicate. So, someone else died. Lots of someones.

Parents. Loved ones. Children.

How many? It's a big number. You're not going to like it.

The lowest I've seen is **30,000 deaths per year**. The highest? **250,000**. This is modern data from the last few years.

The original statistic—the one that launched a global healthcare crisis—was **98,000 per year**. Just in the U.S.

It's similar all over the world.

Worse.

Not every error ends in death. Many result in injury, trauma, permanent harm. Which is even more people.

The damage is staggering.

That's what you're up against. As a leader.

Humans struggle to function as a team. Even when the work is mission-critical. Even when life and death are on the line.

Which is terrible news for us as leaders.

Is your work more important than saving a child's life? Unlikely. Which means you've got a problem

Because I'm guessing you're not in healthcare.

I'm guessing you're reading this to become a better leader. Build unstoppable teams. Run an exceptional business.

First, you need to meet Luis.

Chapter 3

Liaison!

I was a newbie research fellow. A stranger in a strange land.

My days were surreal. A mix of amazing and overwhelming.

When I learned about sentinel events (catastrophic, avoidable mistakes), my brain kind of exploded.

Especially when I learned the root cause.

Do you know what it is? The root cause of all those deaths?

It's stupid. And, I suppose, it makes sense. Now that I know what I know.

The root cause of most medical errors?

A power imbalance.

A difference in status between low-ranking and high-ranking people on a team.

That's the main cause of 98,000 deaths per year.

It's bizarre, right? I mean, I get this in a game studio. I understand why someone didn't mention the little clause about the date.

In healthcare? With professional doctors? Nurses? These are the most educated people on the planet.

Someone must know how to solve this.

★ ★ ★

I went to see Sue. My sponsor. My *liaison*.

Who's the nerd now? Rocking a liaison!

Sue tracked my fellowship. She was kind, well-spoken, with a generous laugh.

"What the heck? One in twenty-five? Is that real?" I asked.

"Yes."

"What are they doing about it? They must be doing something!"

She paused. Then, her eyes sparkled. She had an idea.

"Have you been down to the second floor?" she asked.

"What's on the second floor?"

"Let me show you."

★ ★ ★

Before we take that elevator, ask this question.

How do they train doctors?

Think about it. How do medical schools train doctors, nurses, and assistants?

Not the book learning. Obviously, they study biology, chemistry, anatomy. Bodies. Cadavers.

Not that. I'm talking about the people part. Working with humans.

How do medical schools teach "bedside manner"? How do they teach nurses to stay calm? Be professional with patients?

Because nurses are amazing.

I've been a patient. A husband to a hurt wife. A father to a Make-A-Wish child.

I can't imagine being a nurse.

Patients are cranky. Scared. Sometimes angry. And they're hiding things.

Patients hide things.

Why? Who knows. Embarrassment? Denial? Shame? Whatever the reason, they don't always tell the truth.

So, how does a medical school prepare these excitable medical learners to deal with scared, cranky, complex patients?

Well. Practice, obviously. Learn a skill, then practice it.

And still, the question. How?

You can't let medical students *practice* on real patients. Newbie mistakes? On real people? I think not.

"Huhuh. That looks weird. I wouldn't Google it."

Nope!

★ ★ ★

So, what's the solution?

Consider this. Most doctor visits aren't blood and guts. Or pus spewing. Even when bones are broken, they're rarely sticking out.

In most cases, it's just a conversation.

"I can't poop."
"I can't sleep. I woke up drenched in sweat."
"My penis isn't right."

Not mine. I'm asking for a friend.

In most cases, it's just a doctor asking questions. With a patient sharing details. By talking.

Then, maybe a basic exam. There's poking, prodding, checking of glands. Not all glands. Just some.

If there's something concerning, they order blood work, swabs, imaging.

It's weird when you think about it. Most doctor visits are just two people talking.

Which is great news for medical schools!

They don't need actual sick people. They just need people to *portray* being sick.

★ ★ ★

Enter the simulated patient. Also known as standardized patients.

SPs are incredible.

A mix of actors, teachers, and regular folks. They're smart, empathetic, and oddly skilled at faking a kidney stone.

The SPs at EVMS worked on the second floor. And they didn't just simulate illness. They did something that will matter to you and me.

Because EVMS was a pioneer in this field. One of the top SP programs in the world.

They changed my trajectory as a leader.

Chapter 4

Luis, The Legend

Sue took me to the second floor, where I met Luis—Force of Nature. A master of his domain, he practically invented their SP program.

Luis was impressive. I wanted him to like me.

★ ★ ★

Picture a typical doctor's office. A long hallway. Multiple doors. Plaques on the walls. People moving with purpose.

One exam room per door. Each with a stethoscope, sink, cabinets, and crinkly paper on the big bed.

Like the real thing.

The learner knocks on one of those doors. They could be a doctor in training, a nurse, a surgical tech. Any kind of medical learner.

They walk in to meet their "patient."

Which is an SP playing any number of roles. Young or old. Anxious. Angry. Sitting, pacing. Calm. Sad. Panicked.

The learner interviews their patient. Asks questions. Listens. Pokes. Prods. Touching is allowed and expected.

The learner leaves and enters SOAP notes (the diagnosis and treatment plan).

And there's a twist. The SP has a hidden computer tucked under the table.

Surprise!

While the doctor was assessing the patient, the patient was assessing them right back.

Twenty-one criteria.

Open-ended questions. Agenda setting. Active listening. Recaps. Eye contact. Physical positioning. More.

★ ★ ★

A few minutes later, the medical learner knocks again. Re-enters the room.

This time, the "patient" becomes the teacher. Gives feedback. Offers examples. Allows the learner to practice the tricky moments.

It's one-on-one training for the hardest skill of all.

Communication.

★ ★ ★

I said, "This place is amazing!"

My nervousness was replaced with excitement.

Luis beamed, "Want to sign up?"

"What?"

Another dumb response. First with the admiral. Now with The Legend.

CleverBob SquarePeg, reporting for duty.

He clarified, "Do you want to take our classes?"

"Me? I could learn to be an SP? Is that allowed?"

Luis welcomed me with open arms.

★ ★ ★

I became an SP. Not just in theory. I ran real cases.

Me, a super-nerd, introverted game developer, wearing a flimsy gown, in just my skivvies.

One case had me portraying a Navy SEAL with anxiety. Bouncing legs. Wide eyes. Restlessly pacing the room.

I wasn't *acting*, I was terrified.

And it was worth it.

Becoming an SP made things real. So that the lessons I learned hit home.

Especially the one with the killer death rat.

Chapter 5

Ruthless Clarity, North Star

You and me? We can never beat confusion. Not in a practical, scalable way. Not with a real team of any meaningful size.

Not with creative, dynamic, or complex work. In the real world, you can't eliminate confusion.

★ ★ ★

And we must fight it anyway. Excellence demands it.

So, how? How do we fight confusion?

First, flip it around.

We don't want less confusion. We want the opposite. We want clarity.

Clarity means no gaps in critical knowledge.

It means important details are known by the people who need them, especially during change.

This is your goal. Your north star.

Clarity sounds simple. And it's not. It took me almost a decade to grok how *not* simple it was. How deep this problem really goes.

That's why I use such strong language.

Ruthless clarity.

Ruthless means unflinching. Without compromise. Willing to work through any noise, conflict, or discomfort to achieve clarity.

That's your goal.

And it's going to suck.

★ ★ ★

Games move fast. The whole business is constantly in flux.

Projects. Features. Teams. Shifting budgets and priorities. Plus new hires. And interns—eager-eyed and unskilled.

Teams become imbalanced. Things get out of whack. Sometimes, I have to take drastic measures.

Enter the Team Shuffle.

Shuffling teams is a big deal. I pull all my departments together. Explain the whys, wheres, and hows. The when.

I walk through the details.

Then, I make one thing clear. *Every single person* must answer one simple question. In *writing*.

Question - "Want to shuffle to a new team?"

1) **STRONG NO** - I love my team. Don't move me.
2) **NO** - I like where I am. Prefer not to move.
3) **EITHER** - Happy to stay, or to shuffle.
4) **YES** - I'd like to move, if it makes sense.
5) **STRONG YES** - Please! I need a change.

★ ★ ★

Why would I ask that question? It's super weird. I know. It's not random. I'll walk you through it later.

For now, just remember: clarity is hard.

★ ★ ★

I held a big meeting. I explained the details. I was as crystal clear as it was possible to be.

I stressed, "Every person. Intern to director. *Everyone* must answer the question. In writing."

Then, I opened it up for questions.

Someone asked, "What about me?"

"Yep. Talk to your manager. Pick a number from one to five."

Later, I got, "What about artists?"

"Artists too."

Later still, "What about team leads?"

"Everyone. In writing. Make it official."

Eventually, the questions stopped. We moved on. And I breathed a sigh of relief. I did it. Finally achieved clarity.

I shifted to the "real" planning.

Until I realized I was wrong. One week later. I was talking with one of my managers.

"How's it going? Started talking to your people?" I asked.

I was sure she had. Everyone had started.

"Yep. I got answers from most. No one wants to switch."

"Great! Can you send me the numbers?"

"Um. I didn't get numbers. I just talked to them."

"Did they answer in writing?"

"No."

★ ★ ★

Clarity is so, so, so hard.

With every change. Every pivot. You're moving fast. Launching something new. Even if it's not that new.

Even if you've answered a bazillion questions, something will still slip through the cracks.

You think you've nailed it. And you haven't.

It's kind of irritating, right?

Weren't they listening? How many times do I have to answer the same question?

That crap used to bother younger me. Until I made the realization.

Confusion is everywhere. All the time. It's just a reality of working with groups, even with strong performers.

Our job—as leaders—is to fight it. Create clarity, not just candor.

So, how do you do that?

Chapter 6

Whisper Vs. Tammy

I have a simple mantra. It's critical to our team culture.

Slow Down to Speed Up

This is a great tool for clarity. And it means what it says.

Remember, confusion destroys excellence. In healthcare, it kills people. In creative work, we miss deadlines. We get bad designs. Rework. Critical errors.

Slow down!

Whenever something is new. A pivot. Any change affecting two or more people.

Where there's change, there's confusion. Where there's confusion, there's an opportunity for leaders to shine.

★ ★ ★

Speaking of which.

What is a leader? Most people can't explain it.

It's one of those words we understand intuitively. Like what's obscene. Or good taste. And intuition only gets you so far. Especially when you're talking about your primary craft.

Let's be more specific.

A leader uses influence to achieve results while maintaining trust.

There are three key words. Influence, results, and trust.

Trust is a topic from my other book—*What Makes Great Managers Great*.

Results are a topic for later.

And the word we want now? **Influence**.

★ ★ ★

All leaders have influence. That includes me. And you.

So, what are you doing with your influence?

My suggestion is to use your influence to attack confusion. Because only someone *with influence* can address confusion.

Remember the lesson from med school!

98,000 will die this year from medical errors. Because people on a medical team aren't talking. Even when lives are on the line.

★ ★ ★

There's always a hierarchy at play.

In every meeting. With every action. Someone has the most influence. Someone has the least.

And everyone not at the top of that ladder is afraid. Maybe a little. Maybe a lot.

They will pause. Hesitate. Second guess. Look for the perfect words. All the while, saying nothing, even if they have questions. Even if they know something that might be relevant.

It doesn't matter how much you pay them. Or how much of an expert they are.

Fear is highly motivating. And very real.

★ ★ ★

So, how do surgeons fix this in the operating room? In surgeries? In hospitals?

They don't! That's the whole point. If they could fix it, the deaths would stop.

Instead, they minimize it. Mitigate it with simple techniques.

Like "Time Out."

Billy pees on Johnny in the bathroom? He gets sent to the corner. No talking.

Someone's holding a scalpel over your carotid? Time to hit the pause button. Do introductions. Get everyone in the theatre to speak at least once.

That's a timeout for big kids.

As business leaders, we also have a timeout. Ours comes at the end.

★ ★ ★

It was a typical Monday.

We'd just wrapped up our morning sync. Half the folks left the room. The other half stayed back for a post-conversation.

There was a minor crisis.

We talked it through. Leaders, experts. A bunch of questions. Lots of tangents. Plenty of next steps.

We had a plan. Great!

People started walking out. Chatting. Nodding. Thinking everyone's in sync.

Nope.

Confusion was there. The infected, malicious little rat. *Whisper*. Gnawing at clarity. Hiding in the cracks. Destroying the performance of my team.

Whisper is patient. Singing a lullaby of sweet nothings, "everything's fine." We've done all we can. There's no time.

Screw you, Whisper.

I said, "Hold up a sec. Let's get a recap."

Recap the details of every decision. Include all important points.

This tool is as simple as it gets. No training needed. I just open my mouth and use my influence.

★ ★ ★

If you're not familiar with running recaps, the question you need to ask is, who?

Who should give the recap? You? Your experts? A team lead? One of the decision makers?

No.

Someone with low status should do the recap.

Someone who barely understands what's going on.

Like Intern Tammy. She's only been here a few weeks. Green as grass in the spring

"Tammy? Can you give a recap?"

Tammy's a newbie. She's confused. Scared.

Which is great. Not because we're mean. It's great because she'll mess up.

When she does? Someone will correct her. Because Tammy isn't threatening.

"Back up a sec, Tammy. The release candidate is Tuesday. The launch is Friday."

"Right. Sorry."

"No worries. That's why we do a recap."

Your experts aren't afraid of Tammy. Plus, people like to look smart. Prove they were paying attention.

And believe it.

You start doing recaps? Expecting someone to recount *all* important details? People start paying attention.

Take that to the bank.

★ ★ ★

Also!

Tammy isn't fluent with the lingo. Doesn't know the secret handshakes. The acronyms.

That's a good thing. Because jargon creates confusion.

Like earlier.

I used the word "grok." Maybe you knew the definition, maybe you didn't.

Grok means understanding something fully, completely, deeply. It's from *Stranger in a Strange Land,* an old sci-fi book.

Using grok in a sentence is me strutting my fat butt on a runway. Flexing my street cred. A bona fide, classically trained nerd.

With a liaison!

Makes me look smart. Feel elite. And none of that matters! The goal is clarity.

Minimize jargon and acronyms.

People use jargon because it lets smart people talk faster. Or feel elite.

Whisper loves the taste of jargon, sauteed in butter with onions and thinly sliced red peppers.

★ ★ ★

These tools aren't rocket science.

It's grade-school stuff. Without the fluffy, wuffy. And there's no binkies for nap time.

Your team's not going to break their routines to slow down. They won't spontaneously start doing recaps. Or reduce jargon.

The fastest way to kill a horse? Put two people in charge of feeding it.

If you want to attack Whisper, someone with influence has to step up.

That means *you*.

★ ★ ★

Here's one last tip before we unlock our bunker buster.

Get more voices in your meetings.

Your extroverts love speaking in meetings. Your self-assured folks are fine too. These are your card carrying members of Camp Talkalot.

It's the rest of your staff that are challenging. The quiet ones. The thinkers.

You need them to become part of the conversation.

Every additional voice *now* helps reduce confusion *later*. It also affects other things, like engagement. Which is a great topic. Much less death and no rats.

We'll get to engagement soon enough.

For now, let's finish clarity with a big fat bazooka.

Chapter 7

Thoughts, Questions, Concerns

The SP educators gave me a doozy of a case.

Me, the nerd. Portraying one of the most elite warriors on the planet—a Navy SEAL.

Only this time, the learners weren't med students.

I'd be working with real, practicing navy doctors. Docs who'd just come back into port. Doctors who spent their shore duty getting refresher training.

At med school.

★ ★ ★

I sat in the flimsy robe, in my realistically fake exam room. The doctor asked me questions—lots of 'em.

"Are you smoking?"

"No."

"Doing your PT?"

Physical training. Exercise.

"Of course."

"Do you drink?"

"On weekends. Sometimes."

That was a lie. I was told which things I should or shouldn't lie about. I had a detailed backstory.

The SEAL loved his job. Had to get back to his unit. And he was having serious medical problems. Bad enough to see a doc.

"Shortness of breath?"

"No."

"Caffeine?"

"Sure."

The doctor peppered me, rapid fire, with a bazillion questions. Then he listened to my heart. Did some poking and prodding.

Not that kind of prodding. It was an anxiety case. Not a prostate check.

Speaking of which, how do you think they teach rectal exams? Gross, right? And someone has to teach it.

If you guessed SPs, you'd be right. Prostate glands, kidney stones, anxiety. Whatever is required. They're amazing!

The doctor finished his non-prostate exam. Then, he left the room to submit his diagnosis.

I opened the secret computer. I graded the doctor's performance. Noted which techniques he did or didn't use.

When he returned, I was ready.

No longer the anxiety-ridden SEAL. I was now a trainer. A teacher.

★ ★ ★

It was my turn to ask questions.

I asked, "How do you think it went?"

"You nailed it. They're afraid I'll keep 'em stateside. They don't want to talk to me."

Patients hide things. They keep secrets, like employees.

I said, "There's a technique we could try. Want to give it a go?"

"Anything. Yes!"

"This guy's afraid, either way. If you keep him from deploying, he fails his unit. If he has an 'incident' in the field? Dizziness, palpitations? He gets someone killed."

"Okay."

"He'll talk. *If* you ask the right questions."

★ ★ ★

Before we can get to the right questions, here's two wrong ones.

Direct questions: have a yes/no answer.
Focused questions: have one word answers.

Those were the kinds of questions the doctor asked. Fast. Efficient. Terrible.

Direct questions limit thinking. They make people myopic. Plus, they're easy to lie to.

My advice?

Avoid questions with a one-word answer.

★ ★ ★

Employees are like this SEAL.

They have critical information. And we're going to have to work to get it out of them.

Ask open-ended questions.

Questions that require a discussion. A conversation.

"Morning team. I heard there's a new crisis. Where are we?"
"What kinds of problems do we expect?"
"What's next?"

Big, broad, open. Fantastic.

Another grade-school technique that isn't.

If it was, then I wouldn't have to teach this so often. I wouldn't have to help leaders practice it. In person. Out loud.

Open questions are harder than people think. And they're critical. Because they close the gaps of knowledge. The cracks where Whisper hides.

Use open-ended questions as much as possible. Master them, if you can.

It's a superpower.

★ ★ ★

The funny thing about open-ended questions? Most of them aren't actually questions.

"Someone catch me up."

It's an implied question. Like these three words:

"Tell me more."

Some of the best words you will ever use. Right up there with, "I love you."

"BigMegaCorp rejected the build!"

"Tell me more."

Lovely.

★ ★ ★

That's the basics. Now, the bunker-buster.

"What thoughts, questions, concerns?"

This question works everywhere. In one-on-ones. Small groups. Even my biggest meetings, with multiple departments.

"What thoughts, questions, concerns?"

So much better than "any questions?" That's the question we were taught in leadership seminars.

"That wraps up our presentation. Any questions?"

They taught us the wrong technique. Don't ask that.

"Yeah, but Curtiss. I'm asking for questions."

Time Out!

"Yeah, but" is for newbs. For know-it-all, nerdy engineers with a bag full of Dungeons and Dragons books.

Don't be younger me, Captain ButsALot.

Be older me, Inspector NoCracks.

★ ★ ★

"Any questions?" doesn't accomplish what we think it does. It asks, "Do you *have* any questions?"

That's different.

It creates this weird space where people wonder if their question is *worth asking*.

Is my question clever? Important enough to ask in front of all these smart people? Or does it make me look like I wasn't paying attention?

None of that self-doubt matters to the 98,000!

If it's a struggle when lives are on the line, then it's a struggle for us too.

Interns. Directors. Individual contributors. Executives. Newbies. Professors.

Everyone struggles with psychological safety.

That's the belief that you won't be rejected, embarrassed, or punished for sharing thoughts, making mistakes, or asking questions.

Everyone has fear. And "any questions?" triggers that fear.

★ ★ ★

Do you ever find questions kind of tiring?

Not the first one, obviously. Or the second. I'm talking about the fifth, sixth, and seventh questions.

You've felt that, right? Be honest. Leader to leader.

Were we *really* expecting lots of questions? When we asked, "Any questions?"

That's why the content goes right to the end. A thirty-minute meeting, with ninety seconds for questions.

Mr. Oopsie OutOfTime.

You held a big meeting with hundreds of people. Then you put up a slide with "Any Questions?"

Yeah right.

Whisper loves those token gestures.

I hire smart people. They've figured out the "any questions?" game. Your people are smart too.

Even the troopers from Camp Talkalot. They're gonna pause before they start blabbing in a large group. They're going to make sure they have a dang good question.

Which defeats the point. Confusion isn't about cleverness. Or finding the right question.

Confusion is about gaps.

Whisper is everywhere, creating cracks. Whether you have three people or three hundred.

It doesn't require careful thought to find a gap. We just gotta overcome the fear.

What thoughts, questions, concerns?

That phrasing leans into failure. It implies I made mistakes. Me. The speaker with the grand plan. It implies I rushed. Or mixed up words. Glossed over details. Lost people somewhere.

It attacks confusion directly. It's ruthless clarity.

Whisper hates it.

★ ★ ★

"Thoughts, questions, concerns?"

That's the Bazooka of Clarity. A powerful, enchanted weapon with unlimited ammo.

And don't just rush past it. Ask it. Sit there. And wait. Look around. Pausing. Ten seconds. More!

Pause until it's uncomfortable.

People will get fidgety. They'll look around too. If you still get nothing, address it directly.

"Look, it was a lot of material. I know you have questions. Thoughts. Concerns. Let's hear 'em."

You'll eventually get questions. And when you do, push for more.

"What else?"

Keep asking. Until the well runs dry.

★ ★ ★

You don't have the time?

I hear you. Leaders are always rushed.

My teams release four hundred builds a year. Tens of millions of active players. We have critical pivots almost every week.

Time is always against us.

And the bigger the crisis? The more serious we get about that annoying rat.

We slow down *now* to speed up *later*. We make time for recaps. Ask big, broad questions.

Because when I don't have time for these things? Then being exceptional is the last of my worries.

I'll be leading my teams toward mediocrity. Like Salieri from the movie *Amadeus*. A portrayal of jealousy in the face of Mozart's genius.

Salieri declared himself "the patron saint of mediocrity."

And I do not accept his absolution.

★ ★ ★

So, how did my portrayals go? I played the SEAL. I taught the doctors as best I could. It was stressful. Intense.

Then, it was over. I sat in my exam room until the announcement came from the ceiling.

"SPs, the learners are done. You can dress and leave your rooms."

I pulled on my slacks. Shirt. Tie. Reviewed myself in the mirror. The Research Fellow—a different kind of portrayal.

I turned off the lights and left.

I walked the hallway, lost in thought. Until someone called out.

"There he is!"

"That's the guy!"

Two of the navy doctors were sitting in an alcove.

"We were just talking."

"He's the one. The open-questions guy."

"That's great stuff. I'm gonna use it with my wife!"

We had a laugh. I walked on, smiling. Aware that something profound had just happened.

I didn't grok the significance at the time.

Not until Chapter 14.

Chapter 8

The Spectral Elephant

So that's the first pillar. Now, the second.

Strong performers.

This is an awesome topic. Less death and no rats.

Lots of good tips in this section. And before we get to the good stuff, we need to address an elephant.

★ ★ ★

Not long ago, on a typical workday, there was a knock on my office door.

"Is now a good time?"

Joe was one of my engineers.

He was gentle. Likable, with a good attitude. And unfortunately, he wasn't performing very well. He was on a performance improvement plan. A PIP.

"Hey Joe. How can I help you?"

Joe walked in with more confidence than I was used to seeing. He had practiced for this moment.

He extended his hand. "Just wanted to say goodbye."

Wait. Goodbye?

Did he already know the outcome of the PIP?

We'd been working with Joe for a while. Coaching him. Giving clear feedback.

Guess he figured it out. Good for him.

I replied, lamely.

"Um."

★ ★ ★

When I started my career, the dinosaurs still ruled the earth. I pedaled up I-95 in my foot-powered sedan, dodging potholes the size of brontosaurus footprints.

Curt Flintstone, rising star in the Stone Age of leaders.

Things were different then.

If I did poor work, I expected to get fired. No ninety-day process. No forms. No lawyerese carved into stone tablets.

The boss just leaned out of his cave. "You're done."

It wasn't harsh or mean.

More like pulling off a sabertooth band-aid. Quick. Clean. Letting both sides move on without torturing people for months.

In many ways, it was better. Even for the employee.

★ ★ ★

I stood there awkwardly, looking at Joe.

I shook his hand.

He said, "I just wanted to say thank you. This is the best place I have ever worked. I appreciate the investment you make. In your people. In me."

"You're welcome, Joe. I'm sorry."

I wanted to say more.

Much more.

★ ★ ★

Low performers are a reality.

You have them. I have them.

Any large group has them. Maybe five per hundred. Ten? Fifteen? Hopefully not that many. Depends on how proactive you've been.

This is the elephant in the room.

An elephant with a scary, spectral cloak. Shrouded in law-yer-speak. Discussed in hushed voices.

The elephant stands in the cubicle aisles. We walk gingerly, lest we bump it. Avoiding eye contact.

Why? Because it's a minefield full of lawsuits.

Managing low performers is complicated. Too complicated to walk through the whole thing again.

If you have a low performer? And you're trying to figure out how to manage that person? I already wrote about that. At length. See part five of *What Makes Great Managers Great*.

For this book, I don't want to talk about *how* to address low performers.

I want to focus on the challenges they bring.

Low performers destroy teams.

Especially small teams. You got a team of eight? If it's got a low performer, it can never be exceptional.

Never.

Which means you cannot avoid this elephant. Or pass the buck. Or wait for someone else to fix it. Not when "good enough" is not good enough.

You need to address low performers.

And while you're at it, stop serving crap sandwiches.

Chapter 9

Crap Sandwiches

Let's step away from the lessons of the '90s.

"Constructive criticism" is crap!

So too is the idea of the crap sandwich. You have heard of that, right? Say something nice. Hit 'em with a hammer. Then, nice again. Good-bad-good.

These are garbage concepts.

Think about it. You ever go to your wife? Husband. Friend. Partner. You're having a meaningful conversation. A heart-to-heart.

You say, "I have some *constructive criticism* for you."

BZZZT!

Wrong answer. That's a one-way ticket to couch-ville.

★ ★ ★

Constructive criticism is just criticism with a fancy name.

It's a shortcut. An "easy way" out, masquerading as wisdom.

It was invented with a promise. To give leaders faster control. Allow them to assert their authority while still feeling sophisticated.

"I don't understand why you're so upset, Joe. It's constructive criticism."

What a crock. Those crap techniques were doomed from the start because they were built on pillars of sand.

Ask this.

Do we give constructive criticism to our boss? To the CEO? The board of directors?

I doubt it.

Do we serve up crap sandwiches to our partners? Friends? Customers? People we love?

Not likely.

Which begs the question.

Who is the target of constructive criticism? Who gets to eat those crap sandwiches?

The answer is: people *below* you. Employees with *lower* status.

Constructive criticism is a tool. Only used. When there's an imbalance of power.

BZZZT!

These are crap techniques. Invented in the age of the dinosaurs. And they smell like brontosaurus dung.

Here's my advice.

Stop giving constructive criticism.

There are no shortcuts on the path to excellence, even with pocket rockets.

Chapter 10

Pocket Rockets and Performers

It was a beautiful day outside the office. Inside, I sat at my square table. My boss sat across from me.

She said, "In your self-review, you wrote about your teams."

It was quarterly-review time.

My boss was referring to the big document I'd written, covering everything I'd worked on over the last three months.

She continued. "When you write about your teams. You talk about things like world-class. High performance. Exceptional."

She gave me a thoughtful stare.

"How do you know if you've achieved that?"

★ ★ ★

Great question.

And she was right. I do use that kind of language. World-class. Exceptional.

I have insanely high standards. Mostly because I work in a brutal industry.

Also, because I'm a maximizer. I love taking things from good to great. Pushing toward excellence. In me and my teams.

It's a requirement and an obsession.

★ ★ ★

My boss was sort of calling me out. She wanted to know how I measured team performance.

I suspect that most leaders would answer with a bunch of key performance indicators—KPIs.

There are lots of KPIs.

Player retention. Games played. Time to market. Cost of development. Marketing fees. Bugs. User acquisition. Profitability.

You can use KPIs for games. Or features. Or even for a whole company, year-over-year.

KPIs are good stuff.

The question is: do they apply to teams? Can you assess team performance with a run-of-the-mill KPI?

I used to think I could. Especially when one of my teams had a great year.

"Check out those KPIs! I'm an awesome leader, right?"

As my purview expanded, I saw larger organizations with more complex challenges.

Over time, I realized I was naive. Normal KPIs can't be used to measure team performance.

★ ★ ★

Here's the problem.

There are no guarantees with creative work.

Games. Entertainment. Research. Or any kind of product development.

Whether you have a good team or an exceptional team, there's no guarantee.

It's like Texas Hold 'Em. The card game you might see on ESPN.

You got dealt two aces—pocket rockets?

Booya!

Then, there's the dreaded 2-7.

Time to fold. Especially at a full table.

And sometimes, Lady Luck intervenes. The person with the 2♥ 7♣ might get lucky. Gets a pair of twos on the flop.

Now they have three twos: 2♠ 2♦ 2♥. Which beats those aces.

Strange things happen on poker night.

★ ★ ★

A bad team is like the 2♥ 7♣. Sometimes, they get lucky.

Sometimes, a terrible studio stumbles blindly upon a vein of gold on a sandy beach. They have a runaway hit. They capture lightning in a bottle.

The flip is also true. An exceptional team—your pocket rockets—they can be unlucky.

Bad timing. Shifting demographics. Rising costs of user-acquisition. A misunderstood nuance. A new competitor.

There are no guarantees when you're developing products. Especially not in the games industry.

★ ★ ★

So, what does that mean for measuring teams? It means simple KPIs won't cut it.

Not unless you have a perfect way to test multiple teams. Side by side, in a controlled experiment.

You'd have to duplicate entire organizations.

Like, maybe if we all lived in The Matrix. Asleep in a world simulated by artificial intelligence—AI. If we lived that way, then the machines could run tests.

Simulate one team vs another with perfect conditions. Figure out the exact KPIs for measuring excellence.

Personally, I suspect it has something to do with donuts.

Meanwhile, we live in the real world.

There are no guarantees. Good is not good enough. Being exceptional is our best shot at success (i.e. expected value).

★ ★ ★

I didn't need to explain all that to my boss. She already knew.

That's why she asked the question in the first place. She wasn't really calling me out. She just wanted to understand my thinking.

I said, "There are three things."

I stuck out my thumb. Counting.

"First, communication. The stuff we talk about all the time. Confusion. Collaboration. Working as a team."

I paused. Pointed to my index finger.

"Strong performers. It's easier to flip it. See what percentage are low performers. That's gotta be small. The rest are strong performers."

I stuck out my middle finger.

Out, not up!

"And they need to be highly engaged. Minimum sevens. Eights or nines, preferably. Out of ten. Highly engaged."

"The goal is highly engaged, strong performers who communicate well."

She nodded, accepting my logic.

★ ★ ★

Up above, I said something weird.

"See what percentage are low performers."

It's an odd way of looking at it. Right? I could have focused on all the superstars. Instead, I was focused on that big scary elephant.

The percentage of low performers.

★ ★ ★

Every sizable company has low performers. It's maybe four per hundred. Or six. Eight. Ten at the worst.

You have more than ten low performers in a hundred? That elephant's kicking your butt.

You're reading the wrong book.

Go fix that. Then come back.

As leaders, we gotta get that number down. Work it aggressively. Drop it as much as we can. Eight, six, four.

Three is ambitious. You have just three low performers in a hundred?

You've won the jackpot. Congrats! Sleep well at night, Roxxor Leaderbomb.

★ ★ ★

Let's assume we're doing our best. You and me. As leaders.

We're looking that elephant in the eye. Undeterred by that spectral cloak. We're addressing low performers.

Now, the big question.

Who's left?

If I've got five low performers. Who are the other ninety-five? Who are those people in the middle of the bell curve?

The ones most companies label as "meets expectations." The so-called "average" workers.

Who are those people? I'll tell you.

They're the ones making all the money. They do the work. Design the features. Write the code. Create the art. Fix the bugs. Ship the games.

They are the business. Full stop.

Said simpler.

They are strong performers.

★ ★ ★

That last sentence is important. It's a radically different perspective.

If we only have five low performers in a hundred.

And everyone who's not a low performer is a strong performer? And even better, many will also be star performers.

Then that means:

Almost all our folks are strong performers.

Consistently adding value, helping the team succeed, and meeting the requirements for their title.

93%? 95%? 97% if you're crushing it.

They're all strong performers. At the very least.

★ ★ ★

That's not what I learned in leadership training. They talked about constructive criticism. And crap sandwiches.

The Flintstone leadership tablets lumped it all together. One set of tools for your whole group.

And in so doing, they put strong performers and low performers in the same category. Then they dressed it up with fancy labels.

"Hey, Bobbie GreenShoots! It's constructive criticism. Good-bad-good. Nifty, right?"

Get your dinosaur out of here!

If 95% of my staff are strong performers. If they are running my business. Generating value. Then those other techniques make no sense.

If you're sitting across from a strong performer. Conducting a quarterly review. Going over their work.

Then why would you criticize them? Why make them eat a good-bad-good, crap sandwich?

It's bad leadership, based on a pillar of sand. A shortcut that's only possible because there's an imbalance of power.

There are better ways.

Chapter 11

The Other Influence

If the vast majority of my people are strong performers, then I need different tools.

New ways of thinking.

Like this gem.

Control is not an option.

Firstly, because people are not machines. Not yet, anyway. Not until the robot overlords take over. The so-called singularity, which is hopefully never.

Until then, we are still working with people. Who can't be controlled.

Secondly, because control is not the path toward excellence.

I don't want mindless automatons dancing when I pull the strings. That's a puppet show. It's lame, unless we're teaching kindergarten. In which case, dance away!

You got people sitting around? Waiting for someone to pull the strings? Be told what to do next?

"Salieri's on the line. He said something about ... ab-so-lut-something or other."

Mediocrity in action.

★ ★ ★

Control is a four-letter word. Get rid of it. Replace it with "influence."

Influence is how leaders lead.

It's our bread and butter. Meat and potatoes. The lemon meringue pie, made at home, with a fresh crust, and fluffy, melt-in-your-mouth meringue, cresting four inches high!

Like Mom made at holidays.

Thanks, Mom.

There are a thousand ways to influence others. And make no mistake. Influence is about other people. Our ability to impact them.

Which means we need to address another elephant.

Manipulation.

We love to hate our sneaky villains, don't we?

Lord Voldemort from *Harry Potter*. Scar from *The Lion King*. Senator Palpatine from *Star Wars*.

Spoiler alert—Palpatine was a Sith Lord.

Whispering in the dark. Tricking. Deceiving.

And I use those words on purpose. So we can face the second elephant.

Influence is a sister of manipulation.

If you can influence, you can manipulate.

I've studied sociology, psychology, and player motivation. I know the words to say. How to say them.

I've practiced for decades.

If this was a video game, I'd have a little meter showing my influence skill. It would be quite high.

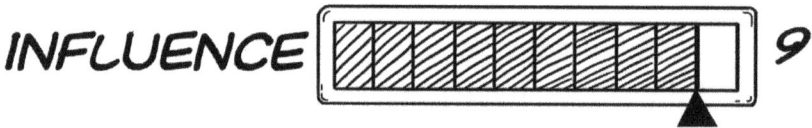

INFLUENCE ▨▨▨▨▨▨▨▨▨ 9

Which means I can get people to do things. I can change opinions. Steer folks in different directions.

The techniques I'm sharing in this book. They will give you that power, too. The power to influence others.

Influence and manipulation are two sides of the same coin. The difference is intent.

If you're tricking people? Lying. Deceiving. Twisting. Getting them to act against their own best interest. Then you're a villain from a Disney movie. A caricature of evil.

The Velvet Tyrant. Worse than Whisper!

Don't be velvety.

Instead, use your influence to help people become stronger. Increase their engagement. Remove confusion. Increase productivity.

Help people achieve the impossible. Become unstoppable. In the best interests of themselves, the business, and the customer.

If you're doing that. We don't call it manipulation.

We call it leadership.

Chapter 12

The Right Kind of Praise

We're talking about strong performers.

Specifically, what we can do with our influence to help our strong performers become even stronger.

Let's get some more meat. The most powerful weapon in the arsenal for strong performers: the Bazooka of Feedback.

Praise the behaviors you want repeated.

That's your ticket to the Awesome Train.

And it's backed by science. The laws of learning, the five-to-one rule, the growth mindset, and the incentive theory of motivation.

If you want to know the why—the details—it's in the *Great Managers* book.

For now, don't worry about *why* this works. Just know it works.

★ ★ ★

"Great job. You were awesome today!"

BZZZT!

Even average leaders know they're supposed to praise their people.

We've all heard that. Unfortunately, no one taught us to do it right.

There's a wrong way to praise—and it's the way most people do it.

"You were awesome" is a judgment of the person.

It's you, judge and jury, bestowing the mantle of Captain Awesome. Might as well slap on a little badge and a cape.

Stick 'em up on a pedestal.

Most of the praise that people give is judgmental. And even though it's positive judgment, it still causes harm.

Compliments can be crap.

Even though they are well-meaning, they can still cause harm, reduce creative thinking, and reinforce a fixed mindset, inhibiting growth.

★ ★ ★

Think I'm making much ado of nothing? Then, ask yourself a question.

What is an employee supposed to do when you give simple, judgmental praise?

Think about it.

You tell your superstar, "You were awesome!"

What do they do with that information? How do they earn that acclaim again? What should they do to get you to say it next time?

Cause once you bestow upon them the grandest of titles, *Princess AwesomeSauce of Fantastic Fields*—no one wants to be dethroned.

They'll need to hear it again.

So, what should they do? How do they get you to call them "awesome" in the future?

It's not clear.

Which means they will start guessing. They'll wonder.

"Was it those pictures I added? The boss likes graphics. Or my dad jokes? My new digs? I dressed up today."

Those aren't the reasons. Well, maybe the dad jokes. Everyone loves dad jokes.

Why did the scarecrow win an award? He was outstanding in his field.

Humor's not my strong suit. I'm more the quirky professor type. "Roads? Where we're going, we don't need roads."

Don't let your strong performers guess the destination.

Point them in the right direction. The express train toward Victory Junction.

That's an actual place—south of Raleigh, North Carolina. A special place with a magical lake, the Kiss & Release Marina.

★ ★ ★

"Wow! You gave a strong intro, covered the details, and guided the conversation toward a clear decision. You fostered good discussion. Engaged the clients. And somehow, finished five minutes early. Nice work."

Mic drop!

So many behaviors were highlighted at once. Fantastic!

And yes. It does take longer to give great praise.

Don't worry about how long it takes. If you're praising behaviors. Doing it right? Then take as long as you like.

It's pure gold. More please!

★ ★ ★

Being praised by your boss? Or a leader. Or someone you respect. It can be one of the most motivating forces in the universe.

So, don't waste it!

"Thank you for the detailed exploration. Your quick thoughtfulness helped us make a decision, without slowing the rest of the team."

Translates to:

"My boss likes when I'm detailed. Prefers quick responses. Wants solutions that don't slow others."

Your strong performers will notice. They will begin to repeat those behaviors.

And if you skill this up?

Imagine using this technique every time you open your mouth. Praising the right way. Imagine a world where you almost never highlight the weaknesses of your strong performers.

Imagine never using constructive criticism again.

You're mastering these techniques. Moving mountains, without criticism. Building an unstoppable team.

And you're not done.

You keep practicing, eyes on the prize. You learn to praise entire groups.

You're in a meeting. Could be big, small. You see Sally leading a discussion. She asks a great open question.

Sally says, "What thoughts, questions, concerns?"

And you notice it, in real time!

"Nice open question, Sally. And she's right to ask. What thoughts, questions, concerns do we got? Come on, we know there's confusion. Bring it!"

Others will start asking that question too, even when you're not around.

You can praise a recap. Or iterating without attachment. Or whatever behaviors lead to success in your world.

★ ★ ★

Once they get good at those behaviors? Switch it up.

Find new behaviors to praise. The critical pillars that get the results you want. Whatever leads to success.

It's up to you.

Thoughtfulness. Attention to detail. Collaboration. Openness. Speed. Proactivity. Engagement. Focus. Intensity.

The application of a skill. Upgrades. Personal growth.

There's no end to what you can praise. As long as you know what you want.

That's your job. Figure out which behaviors lead to success. For both individuals and teams. Then praise those behaviors.

Again, and again, and again.

This is the Bazooka of Feedback. The express train to Crush-It City.

And here's the best part. You can start using this technique today.

It's easy, unlike kiteboarding.

Chapter 13

Snowboards and Surfboards and Kiteboards, Oh My!

Remember the early days of COVID?

Everything was shut down. Whole companies went full-remote. At least out here in Southern California. It all shut down. And it went on *forever*.

One day, I was talking with Macy over Zoom. She was a director in another part of the studio.

I asked, "Where are you now?"

"Hawaii."

"How long are you staying?"

"Forever."

"They let you move?"

"I don't know. I didn't ask. I'll never have a chance like this. Now, I Zoom in the mornings. And surf in the evening. Every day, living the dream!"

"What if they make you come back into the office?"

She shrugged.

"Guess I'll get a new job."

Which she did.

★ ★ ★

Macy was crazy. No, not crazy. Ballsy!

She inspired me.

What if I moved temporarily? Obviously, not to Hawaii. I loved a different kind of board sport.

I set my sights on Utah.

Ten months to plan—COVID kept on trucking. Plenty of time to get in shape for snowboarding.

My wife and I rented the first floor of an old home. We piled in our beat-up Highlander. Candy, our spoiled mutt, slept on Jenny's feet.

Nine hours to the base of the Wasatch Mountains, where we spent seven weeks near Brighton Resort.

I lived The Dream™.

Up early each morning. First in the lift lines. From The Crest to Thunder Road to Snake Creek Express.

That's where my joy was. The trees around the Pioneer and Sunshine trails.

The crisp air. Just me and the soft crunch of powder on the untouched twists beneath the tall evergreens.

Alone, in a good way. It was a once-in-a-lifetime experience. And then it was over.

We piled back in the truck and headed home. And during one of those long, quiet stretches, I began to wonder.

"What should I do now? I'm in the best shape of my life. Don't want to waste it."

My wife suggested, "Skateboarding?"

"If I were younger."

"Surfing. That's what everyone recommends."

Hmmm. Not a bad idea. I should have thought about it.

★ ★ ★

One summer, when I was twelve.

I tagged along with my older brother and his much older friends. They gave me a beat-up board. Then paddled off.

I was all by myself on the waves. Alone in a bad way.

No wetsuit. Poor vision. I could barely see anything. With no idea what I was doing.

God! It was cold.

I ended up lying on the sand, almost in tears, shivering under a flimsy towel.

Sand Solo Sidekick.

Not the way to fall in love with a sport.

★ ★ ★

"No," I told her. "No surfing."

"What about kiteboarding?"

"What?"

She held up her phone, and I stole a glance. My mind was blown.

A board on your feet. On *top* of the water. Pulled by a massive kite, up in the air.

It looked technical. Challenging. Ballsy.

Yes, please!

★ ★ ★

When I was younger, we used to hike up Mount Trashmore. That's a real place where the best kite-flyers did their thing. The mega-hobbyists.

They had fancy kites. Multiple strings, doing loop-de-loops in the air, like a ballerina.

Beautiful!

Now take that kite. Multiply it by 10! Expand it until it's forty-five feet wide. That's 15m for those in the sport. Practically a small airplane.

Now, attach that kite with seventy-five feet of the strongest string on the planet. Then lock it to your chest with a steel carabiner.

With that much kite, the human becomes a puppet on strings. And the tyrant is the wind.

That's the premise of kiteboarding.

You paddle into the ocean. One hand is guiding the 15m sail-cloth, 25m above your head. Swooping it, back and forth.

While you're swimming! With only your feet.

Because your other hand is busy too. It's holding the board itself. A kiteboard that's a bit like a snowboard. Slightly fatter, shorter, and more square.

You're doing three things at once.

Paddling against the waves, slapping you in the face. Holding the kiteboard, trying to keep your body perpendicular to the shore. And steering a monster of a kite, swooping it back and forth, lest it crash into the ocean.

Then, you go. You have three seconds to swoop the kite, attach the board to your feet, and stand up.

Who invented this crap?!? It took months of practice and training. Until one day, I did it!

Whoosh! I was a kiteboarder!

I was also exhausted. I paddled in. Shared the excitement with the love of my life. And uttered the infamous phrase.

"Just one more run."

Winds don't announce they're changing. The Sky Tyrant blows harder, high up in the air where you can't feel it.

I tried to launch, like normal. A gust caught my kite and sent me flying through the air like a discarded toy.

Thirty feet … Kersplat! Into unforgiving sand.

I heard the crack. I felt the snap of the bones in my left arm. Multiple bones, the worst pain of my life.

I was told that scars were cool. I'm not sure my wife agreed.

★ ★ ★

This is my sad COVID story.

And the point is hopefully obvious. Let's spell it out anyway. For clarity.

I was a kick-butt snowboarder. Seven weeks in the Utah trees!

And I couldn't kiteboard for crap.

It's a lesson as old as time. And it applies perfectly to our strong performers.

Your strong performers. Your star performers. They're not equally good at all things. So, quit treating them like they are!

Empower the skill—not the person.

Which skill? Every skill. One at a time.

Coding. Debugging. Design. Art. Illustration. Animations. Shading. Leadership. Research. Testing. Working with customers. Communicating. Speaking in front of a group. Writing a proposal. Architecting whole systems. Optimizing code. Detail work.

Anything. Everything.

There are a bazillion skills. And even our very, very best. Our star employees.

Just because they can snowboard, doesn't mean they can kiteboard. Or surf.

Even if all the skills involve water and a board on your feet, they're not the same.

Don't empower the person. Empower the skills that they have mastered.

★ ★ ★

Think about your best people.

You know who they are. The ones you give your toughest challenges to. The stars!

That's how we think of them, right? Rockstars. Superstars. Megastars.

We'd clone them if we could. Give them fancy names. Duke of Deadlines. The Fixinator. The Duchess of Done-Done!

We love their competence. Skill. Ability to pull a clutch victory.

That love pushes us to be bad leaders. Setting our strong performers up for failure.

It's like a mini-version of the Peter Principle—promoting people to their level of incompetence.

We start empowering them everywhere. Instead of where they are best.

Remember the basics—everyone has limits. You, me. Your stars.

Know their limits.

Know which skills they've mastered. Which are progressing nicely. And which are still nascent, just out of the womb.

★ ★ ★

"Wait a second, Curtiss! What about the weak spots? The stuff they suck at?"

This is where you come in.

Figure out which skills you want them to grow. Better yet, *ask them* which skills *they* want to grow.

Then, put on your thinking cap. Create opportunities that align with those growth areas.

Find harder challenges. Not too hard. Just the right amount of hard. The sweet spot.

Be a leader. Keep your eyes open, look for special projects with ideal growth conditions, person by person.

Finding those will take time. And thoughtful leadership.

Cause once you've found the perfect challenge, the rest is simple.

★ ★ ★

"I have an opportunity for you. It's outside your comfort zone. It's a good skill for you to master. And don't worry. I'm gonna coach you through it as you skill up."

No sneakiness required. No raised eyebrows or secret villain laugh. Mwuahahahahah!

Just be direct.

Tell them it'll be hard. Tell them it's for their own good. For their growth in their career. And tell them you'll be helping them through it.

Make it clear that you'll be directly involved.

This sets you up for later. When they are trying, practicing, and making mistakes. You've already told them in advance to expect that.

They'll be ready for you to offer some direct, hands-on feedback. Specific, detailed. As much as you need. In this state, people won't feel *criticized*.

They'll feel *guided*.

It's how you avoid the constructive-criticism trap.

Someone's learning a new skill? They expect help. They're less defensive. More open, willing to listen.

Lead. Refine. Repeat.

Just make sure you tell them what's happening. Ruthless clarity!

Don't empower folks on the hard challenges. Tell them the opposite.

"You're not empowered on this. I'll be helping you with this."

Use the four stages of situational leadership.

Supporting, coaching, directing. Depending on their level of strength.

SITUATIONAL LEADERSHIP

DIRECTING   COACHING   SUPPORTING   EMPOWERING

Leading strong performers through these stages was awkward for me at first.

A younger me tried to be clever about it. Used careful language, just at the edge of being sneaky.

What a dork.

In my cleverness, I created confusion. Resentment. I messed up a few of my folks. I learned the hard way that it's better to be direct.

Be up-front with your intentions.

Simpler, with less stress.

"It's gonna be hard. Not like your other tasks. Let's check in daily. I'll offer suggestions, provide guidance. Especially in the beginning. As you get stronger, you can start taking the reins."

Simple. Direct.

Which allows you to focus on the hard part.

Separating out the rock-solid skills from the weak ones. Deciding which are ripe for growth.

Figure out what they themselves are motivated to learn, then coach them until they're rocking that too!

And speaking of coaching, let's talk about that.

Chapter 14

Questions Are The Answer

Coaching at work isn't like coaching on the soccer field. That's a story for later.

At work, we're dealing with thoughtful, sensitive people. Strong performers who need help, and still don't want to eat a crap sandwich.

People who are challenged because they *need* to be, not because they *want* to be.

Enter coaching.

Coaching is all about questions.

★ ★ ★

Sherise was a hyper-competent contributor. Positive and also quiet. She spoke softly, hesitantly. Especially in public settings. Definitely not the first person you'd think of as a leader.

So, when she volunteered to run one of my core activities, I wasn't sure what to do.

Think about it. I had a highly engaged, star performer. She wanted to tackle something way outside her competency.

There were lots of skill gaps, and she still signed up anyway.

If coaching had a poster, she'd be on it.

★ ★ ★

I asked, "Want to practice your kickoff presentation with me?"

Sherise answered, "Sure. I've been practicing a bunch."

She jumped in. She had draft slides, pictures, and a story.

Unfortunately, her presentation was bad. The material was out of order. The story lacked a hook. There was no galvanizing conclusion. It was a big, boring information dump.

Blech!

What would you do?

You've got this awesome employee. She's pushing herself to become stronger, outside her core skillset.

How do you guide her without crushing that energy?

★ ★ ★

"Okay. You had thoughtful slides. You have good core material there. And you're way ahead of schedule, which leaves time to iterate. Nice."

Everything I said was true.

Even though her presentation was bad, I still found ways to praise her behavior.

Be authentic with praise.

Lies undermine *trust*—one of the three keywords of leadership.

★ ★ ★

Speaking of trust.

Do you trust *me*?

Like that line from the movie, *Aladdin*. He's floating on a magic carpet. He extends his hand to Jasmine.

(That's a great doodle of me. Thanks, Ivano!)

I've been around the block.

I learned critical lessons far too late in my leadership journey. I had to learn how to stop taking shortcuts and start putting in the hard work.

And I appreciate that some of what I'm laying down might seem fluffy to you.

So, I'm asking Aladdin's question.

Trying to show you a whole new world.

★ ★ ★

I had no intention of letting Sherise give a bad presentation.

I also wouldn't set her up for failure.

Nor would I revert to the crap sandwich. Or constructive criticism just because it would have been convenient.

Coaching is usually the right tool.

Careful, clever questions. Guiding Sherise to identify the problems herself.

★ ★ ★

I asked, "So, how do you think it went?"

"Honestly? I'm not sure," said Sherise.

"Tell me more."

"It's missing something. The story sounded better in my head."

"Let's explore that. Tell the story again, in sixty seconds."

I probed her understanding, nudging her to discover the errors by herself. She told her story again. Then I paused. And continued coaching, with questions.

"Okay. So, what are the elements of a good story? Do you remember?"

She recited the three ingredients from memory. That meant she had the basics. She just needed help applying them.

"Great. Now walk through them, side by side with your story."

★ ★ ★

I guided her by zooming in on the story. Probing her understanding, without judgment, until she had The Moment™.

She said, "Wait! I didn't start with a question or challenge."

Eureka!

"How could you fix that?"

"With a challenge! I've got the perfect one. The first time I gave a presentation, I failed. It's a great opener!"

"How would that affect the order of the slides?"

"Yes. I see it. Let me think."

★ ★ ★

Consider this situation from her perspective. She's nervous, engaged, excited, hungry. All at the same time.

She craves feedback.

Not the kind that crushes her soul. She wants thoughtful feedback that keeps her engaged in the growth process, even though it's hard.

Sometimes they'll have a eureka, like Sherise. I love those. Sometimes, they'll get stuck. That's great too.

"I'm not sure."

Wonderful!

Try to get them to hit the limits of their understanding. Admit they don't have an answer. That's the key to unlocking another super-weapon.

Ask permission before giving the answer.

"Would you like some ideas?"

They'll almost always say yes. And if they do, now you can go into teaching mode. Have at it! They're primed, ready to receive.

Educators call this the Law of Readiness. As in, the moment the learner decides they are ready, motivated to learn.

Like me, buying cleats—after I met Adonis.

★ ★ ★

Coaching is dynamic.

It unfolds in real time. It's different for each person, depending on where they are in their journey. Coaching requires me to bring my A-game.

And it's fun!

Seeing that eureka? Seeing them hit the limit of their knowledge? Delicious. And it couples nicely with praising behaviors.

A nice cup of tea with your slice of meringue.

Most of the time, they aren't even aware they've been "corrected."

They leave my office uplifted. Thinking. Excited to go tackle their problems in a new way.

Manipulation? Or leadership!

Let's check the schedule for the hype train. Because as powerful as coaching is, there is a downside.

Coaching takes more time.

You have to ask questions. Give them time to work through it themselves. It's a slow process.

My suggestion? Don't worry about how long it takes. Schedule time to coach your strong performers with questions.

It's gold.

Learn to ask better questions with your strong performers. And with your peers. And partners. Even with those above you in the hierarchy.

Practice until questions become second nature. Until you do it without thinking.

Until you grok a powerful truth.

A truth that started out as a feeling years back. When I was NeckTie FellowBottom at med school, talking with two doctors in an empty hallway.

Questions are the answer.

Eureka!

Chapter 15

Travel Sorceress

It had been a long drive. Ninety minutes in LA traffic.

Which was fine. It was a personal day, so we weren't in a rush.

My wife and I pulled into the parking lot, and immediately ran into a wall of cars. Signs with warnings.

"Tickets required!"

What? Don't you buy tickets at the door? The line of cars crept forward, uncovering more signs.

"Sold out for the day!"

We didn't have tickets, and it was my fault.

"Crap!"

★ ★ ★

A week earlier, my wife said, "Let's do something fun."

"Like what?" I offered.

"I want to go to The Huntington."

"Huntington Beach?"

"No. The Huntington. It's an arboretum."

Not just *an* arboretum.

The Huntington Library, Art Museum, and Botanical Gardens.

It's world-famous.

The library had an original of the Gutenberg Bible. Manuscripts from Thomas Jefferson, Benjamin Franklin, and Mark Twain.

Two quartos of Shakespeare's *Hamlet*! Whatever a quartos is, it sure sounds cool.

That was just the library.

The arboretum was equally famous—120 acres of serene, otherworldly beauty. The gardens often served as the backdrop for our favorite series, *The Good Place*.

So yeah, "an arboretum."

I didn't know any of that at that time. I barely knew what an arboretum was.

What I knew for sure was that Jenny was excited. That was good enough for me—a case where good enough *was* good enough.

★ ★ ★

"Sounds fun," I said.

"Can you help me get tickets? Pick the right times. With travel and all."

"Now? A week out? I'm sure we can get them at the door. It's a museum on a random Thursday."

"I don't know. They recommend tickets."

"They always say that. Let's go watch the game."

What a Chad!

In fairness, I wasn't bro'ing out on her. My wife loves soccer as much as I do. It's a shared passion.

And you know how those artsy sites are.

You make an account. Give your card, address, and email. They want a donation. They start spamming you.

Then you get there to find ten people in the whole place.

★ ★ ★

Which was how I ended up stuck in a line, feeling every inch as the car crawled forward, on a one-way path.

We weren't getting in.

Don't be me. Learn the lesson, without the pain.

Respect your experts.

Jenny is a Travel Sorceress. The Chief Excursion Officer for Murphton Manor.

Thirty-six years of trips.

Lady Gaga in Vegas. Ocean kayaking in Mexico. Hiking the Red Rocks. Yosemite during COVID. My bucket list in Utah.

All the way back to our first getaway: Pink Floyd, live in D.C.

My wife is a master trip planner.

Whereas my claim to fame was making us late to the airport. Missing our first flight as a couple.

And now, my failure at the cave—The Huntington.

★ ★ ★

Why did I second-guess her advice?

Well, let me ask you the same thing. How come you override your experts?

Aw, come on. Don't tell me you don't. We're being honest with each other. I showed you how much of a chump I was.

Now, take a look in that mirror. You, the big, bold leader with Vision™.

★ ★ ★

It's not about vacations. I'm talking about your business. Big decisions. Pivots. Direction.

You start a new initiative. Make a big investment. Acquire some company.

Let's be honest! Not all our crazy schemes are great.

So, what do you do? In those rare moments when someone below you challenges your schemes.

"Excuse me. I have a few concerns," says Sergeant BuzzKill.

That's courage. Challenging the leader's new direction.

★ ★ ★

We say we want our people to raise concerns. Ask questions.

We pay them a lot of money to ensure they speak up. And usually, we like it when they do.

Until it hits too close to home. Threatens a pet project. Pokes holes in our latest dream.

★ ★ ★

General Disregard is a loser.

If your strong performer raises a concern, then you know what?

You should be thanking your lucky stars you have staff who care enough to help you realize your mistake. *Before* you make it!

Then, you should praise their bravery. That's definitely a behavior you want repeated.

You are threatening. I am threatening. 98,000 per year is more than a trend.

Speaking up is scary, and even if the employee isn't afraid, we still need to slow down and listen.

Because they trust in our leadership. They will be easy to sway. They will go along with our plans, even if they have doubts.

Like my wife.

She wasn't afraid. She just wasn't sure the tickets were *absolutely* necessary.

Plus, my enthusiasm was contagious. Dorkzilla FunFactory, reporting for duty.

She thought maybe she was worrying for nothing. All the while, the voice of experience was screaming in her head.

"Los Angeles!!! Twelve *million* of your closest friends."

★ ★ ★

In this story, the pain was nothing.

We missed out on The Huntington. Big deal. The Sorceress had another spell prepared, just in case.

In business? The stakes are high. We ignore our experts at our own peril.

We could be wasting millions. Hundreds of millions.

All because we were more excited about seeing the game. Or didn't have the patience to slog through the details.

Worst of all? Maybe we were drunk on the delicious sweetness of our own designs.

Hear their thoughts, questions, concerns. Think about them carefully.

So you don't get stuck without tickets.

Like a chump.

★ ★ ★

Okay.

That's enough ideas about strong performers. Hopefully, you enjoyed the stories and found some nuggets.

Before we move on, here's one last thought.

Imagine it's the end of the year. Review time.

Try to visualize one of your mid-range performers. Someone with a rating of "Meets Expectations."

Not a star performer. Not exceeding. Not up for a promotion. Just a middle-of-the-road employee, getting an average review.

And this year, something is different.

You've evolved as a leader. You're removing confusion. Thinking about folks as strong performers.

You're praising the behaviors you want repeated. Hearing your experts. Coaching.

And in this scenario—even with an average rating—your employee might say this:

"Those were very nice words. It makes me want to cry. Thank you so much!"

Ruthless clarity.

Strong performers.

And now, let's meet Adonis.

Chapter 16

Adonis

Two down. One to go.

Relentless engagement.

This is where *good* leaders become *great* leaders—where great leaders build unstoppable teams!

★ ★ ★

Many years ago, I got a letter.

> *Dear Mr. and Mrs. Murphy,*
>
> *We're sorry to inform you that we do not have enough coaches this year. If we do not get additional volunteers, then your son will not be able to play soccer.*

Sincerely,
Adonis,
Soccer God.

"Crap!"

How could they not have enough coaches? It was the largest club in southeastern Virginia!

I stood there, brooding. My wife cleared her throat.

She said, "So?"

"So what?"

"*So* ... what are you gonna do?"

"What am *I* gonna do?"

She raised her eyebrow. Tilted her head. Expectant, with a long, pregnant silence—twins, judging by the weight.

"Crap!"

Thus I became a coach of the under eight (U8) boys soccer team.

The Dragons.

★ ★ ★

Adonis! That's how I thought of him. Because when I met him, he looked like a Greek God.

He had the perfect tan. Fit, in the way only seasoned soccer players could be. Every muscle, sculpted.

Crisp digs. Calm. Authoritative.

In a way, he did have godlike powers. Because with some well-placed words, Adonis summoned coaches from the aether. Turned normal, everyday blokes, into Coach Murphy!

It was a tall order.

Nerd central over here. Skater boy. Engineer.

The entirety of my soccer knowledge could be summed up in three sentences. There was a guy named Pelé. There was a movie about Beckham.

And, the rioting freaks in other countries called the game "football."

Which was obviously wrong.

Cause American football? Definitely brings to mind the word "foot." And "ball." The science checks.

"Crap."

★ ★ ★

So yeah, I volunteered. Orange cones. A whistle. Soccer balls.

I scheduled practices, and I was honest with the other parents.

"I know nothing about soccer. Adonis asked. So here I am. I'm committed to showing up. Please be patient."

That's how it started.

I watched some videos. Went to the newbie coaching class. Learned the five-to-one rule from Adonis.

★ ★ ★

Coaching boys was hard!

I needed to be able to run. So, I started jogging. Me! Nerdimus Maximus, pulling hammies tying my shoes.

After a little while, I realized I still had no idea what I was do-ing. So, I signed up for advanced coaching classes.

Soon, I found myself practicing on weekends. Not even sure who's idea that was.

One day, a parent asked me a question.

"Why do you wear sneakers?"

"Me? Cause I'm not playing. I don't need cleats. ... Wait. Do I?"

"You do. You're the coach."

This was a turning point. Buying cleats made me feel official. Armoring up for battle, in my uniform of station.

My intensity increased.

I began watching games for fun. I played intramural soccer with other adults. I began thinking about soccer more and more. I dreamed about coaching.

Months went by. Years. Until one day, I was no longer just Coach Murphy.

I was *Coach*—freaking—*Murphy!*

"Let's go boys!"

★ ★ ★

That's the backstory. Now, let's unpack that third pillar.

Relentless engagement.

Relentless means unwavering. Without pause. Willing to push forward with purpose, no matter how long it takes.

It's a strong word, like **ruthless**, from our first pillar. It's uncompromising, like being exceptional.

The second word—**engagement**. That's less well-defined. If you search it or ask AI, you'll find lots of fluffy definitions.

I don't do fluffy. Here's my simple, practical definition.

Engagement means committed to success.

Nom, nom, nom!

★ ★ ★

When I started out, I told the parents, "I'm committed to showing up."

I gave an oath.

Like the old cliché: 80% of success is showing up.

And that's not what we're after. Right? You and me. Leaders. Trying to get from good to great. From great to exceptional.

My promise to show up was not relentless engagement.

Not even close.

Chapter 17

The Voice of Command

As a coach. Father. Leader. I evolved. From showing up, because I made a promise.

To *Coach—freaking—Murphy*.

My whistle was no longer an adornment. It was an extension of my presence. Often unnecessary, because I'd mastered new skills.

The Voice!

Like *Fus-Roh-Dah*, from the Elder Scrolls games. The voice of command.

Soft or loud. Spreading my influence in shockwaves of positive energy. The Voice compelled, like a great smell. Hot dogs on a grill, wafting over your neighbor's fence.

"Whatcha cooking over there?"

And it wasn't just the Voice.

Everything about me was different. I was confident. Fit. Committed to more than just showing up.

So, why do you think that happened?

Why did I change?

Hint: it wasn't Adonis.

★ ★ ★

Here's the research.

Engagement comes from mastery, autonomy, and purpose.

Three ingredients. Separate. Equally important.

Mastery.
Autonomy.
Purpose.

MAP. Capitalized, because each stands on its own two feet. Probably with cleats. Or maybe boots, for the Brits.

We'll explore each of these, starting with the obvious problem.

Mastery.

★ ★ ★

I'm a game developer. So, pretend I was a character in a video game.

In a game, you'd have little slider bars, going side to side.

You'd be able to allocate points. From zero to ten, where zero is awful and ten is maxed out, practically speaking.

★ ★ ★

What was the mastery score of NoobMaster 5000, with his shiny whistle?

Let's look at the skills.

I was a father. I knew the mischievousness of seven-year-old boys. I could work with parents. Speak in groups.

At the same time, I knew nothing about soccer. Still trying to understand why "offside" was a thing.

Plus, being a father of a young boy didn't make me a coach.

Young boys are distractible. Picking clovers. Chasing plastic bags, floating in the breeze. Goal! For the other team.

Let's allocate some points. And I'll be generous.

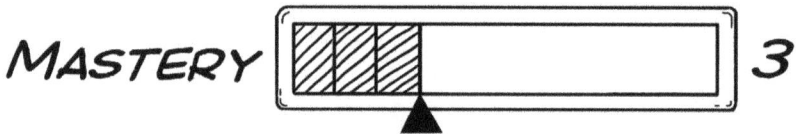

MASTERY ▱▱▱▱▱ **3**

How about autonomy?

This one's not so obvious. Because at first, it seemed like I had tons of autonomy.

I was the coach, right? I volunteered. I was in charge!

Not so fast.

Adonis and his soccer pantheon had "recommendations." When to practice. How to practice. For how long. And where.

Big things. Little things.

Like, not making the boys run laps without a ball. Boys under twelve don't need to work their stamina. They need to touch the ball with their feet.

Who knew?

There were rules for coaching. How to teach. How to praise.

The five-to-one rule.

Which was the genesis of my favorite mantra—praise the behaviors I want repeated.

Thank you, Adonis, Greek God of "*Stop Trying to Build The Next Pelé—Get Them To Fall In Love With the Game!*"

And don't forget the other rules. Players assigned in advance. Games on a tight schedule.

The truth is, I had little autonomy. And that's before you factor in the most important part.

★ ★ ★

Did I want to be a coach?

I'm not asking whether I volunteered. Of course I volunteered. Under the thoughtful suggestion of my wife, who nudged me for my own good.

I'm not asking if coaching was good for me. Or a great way to bond with my son. Both were true.

We're talking about engagement. And autonomy. Which requires personal choice.

I didn't choose this path.

I was minding my business. Probably playing *Oblivion*, one of those Elder Scrolls games.

Fus-Roh-Dah'ing some vampires in a fantasy realm. Instead of melting in the humid, summer sun. Dodging ground bees on a worn field, covered in clovers.

Then in comes God. Or Fate. Or the universe. They said unto me.

"Thou shalt coach! Or thou shalt suffer, from yon wife, The Look."

Adonis needed more acolytes. I was the sucker who said yes. Which is not the same as choosing it for myself.

Added together, not so great.

★ ★ ★

The last factor of engagement is purpose.

AUTONOMY `[▨▨▨        ]` *3*

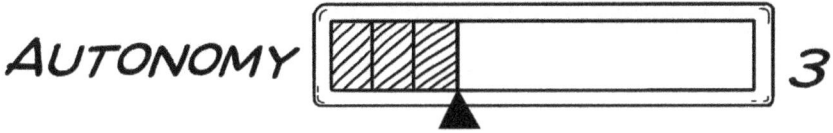

Purpose is the most important part of engagement. And it's also not at all like mastery or autonomy. It's not even measured the same way.

Purpose will require a deeper discussion.

"Let's put a pin in it."

Check me out. Liaison. The Voice of Command. Pin Whisperer!

★ ★ ★

If coaching were a game, my stats kind of sucked.

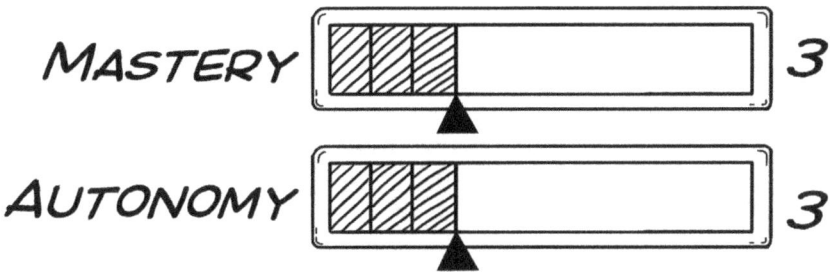

MASTERY `[▨▨▨        ]` *3*

AUTONOMY `[▨▨▨        ]` *3*

Courteous Curt wouldn't be your first pick, in *U8 Soccer Coach Management: the Game.*

Which is weird in hindsight.

When I started as a new coach, I felt committed. I was energetic. Fully intent on showing up every day.

Doing my best.

★ ★ ★

Then time advanced.

I started jogging. Getting in shape. I watched videos. Signed up for advanced classes.

These were personal choices. That's an increase in autonomy.

At the same time, my mastery also improved.

I developed routines. How to start practice. When to scrimmage. Activities for game day.

I unlocked *Fus-Roh-Dah*. Weak at first. I bought my own cleats!

Autonomy shot to a whopping 7!

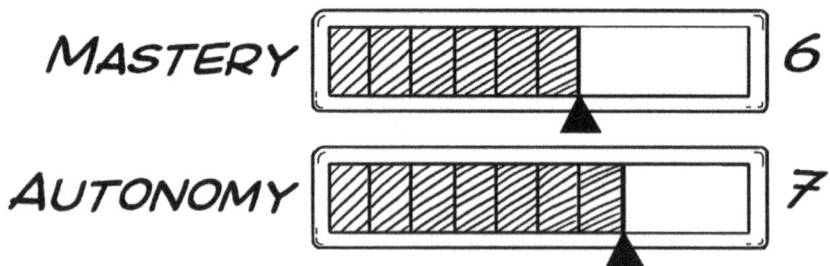

MASTERY   6

AUTONOMY   7

Then, the conversation I will never forget.

"Jacob, it's almost the end of the season."

Jacob was my son. The root cause of this soccer madness.

I asked him, "What about next season? You don't have to keep playing soccer. I know it's hard with me being the coach and all. We can do a different sport."

"It's okay, Dad."

"What do you mean?"

"I like you as coach. It's fun. I want to keep going."

"Fus-Roh-Dah!"

"Fus-Roh-DAAAHHH!"

★ ★ ★

Soccer had two seasons per year: fall and spring. And each time, I asked my son the same thing.

"What about next season? Should we keep going?"

"Yep. Still having fun."

One season became one year. Two years. Five.

My skills leapfrogged.

MASTERY ⬛⬛⬛⬛⬛⬛⬛⬜ **9**

AUTONOMY ⬛⬛⬛⬛⬛⬛⬛⬛ **10**

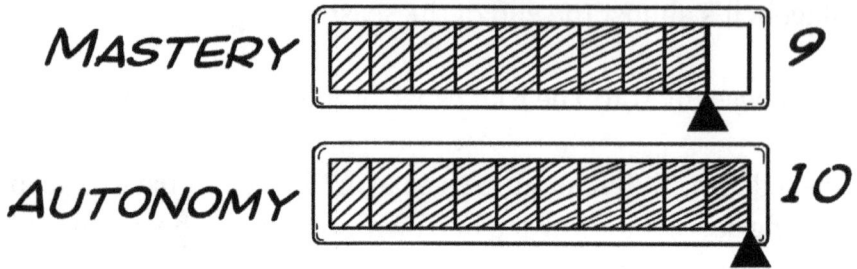

The Goalfather. That's a character you'd pick in my doofy video game.

★ ★ ★

Along with the bump in stats, something else happened. I was no longer just a guy who volunteered.

I was *Coach—freaking—Murphy.*

Coaching became core to my being. Part of my identity. Still is, all these years later.

If you needed me on the field tomorrow, I'd be ready to go. Sure, I'm slower. Fatter, with no cleats.

And if you need me? My *Fus-Roh-Dah* is still maxed out!

Younger me had the maxed Voice, on top of being in shape. He was a rockstar coach of a boys' soccer team.

That's the power of engagement. What we mean when we say, **committed to success**.

Chapter 18

The Thrill of Mastery

Think you got mastery and autonomy all figured out?

Let's see what happens in the real world.

★ ★ ★

I asked, "How's your universe?"

We were sitting at my square table. Maya and I. Having a one-on-one.

Maya was a strong performer. A *very* strong performer, with lots of experience.

My question, "how's your universe?" was nice and open. It got things going. Helped me see how she was doing.

She said, "Well. If I'm honest. I'm a bit overwhelmed."

"Tell me more."

"I feel behind. These last three weeks. I don't have confidence in my solution. I feel … trepidatious. I think that's the word."

We looked it up.

"You're nervous about something that may go wrong?"

"Yes."

Maya had lost mastery. She was worried, concerned about failure.

"Okay. Trepidatious."

Then I paused, thinking about our past conversations. Connecting the dots.

I said, "I'm happy to hear that."

★ ★ ★

Before you judge what I said, consider your ancestors.

Yes. Your ancestors. Way, way, way back. Before history.

Back when a club was the pinnacle of technology. Peanut butter cookies weren't even a dream in Nom-Nomicus' little mind.

There was no Amazon. No insta-delivered "touch-less" meals.

You learned. Or died.

You learned the patterns of animals. How to avoid Yogi the Bear—who was less cutesy and cuddly, more slashy and deathy.

You studied the migration patterns of fish. Which berries were safe. Making a net with reeds.

Life was hard. Learning was how you survived.

So, the brain promoted that. Pushing my ancestor to go learn something new. The thrill of mastery!

"Curtok build fence! More chickens. Nom, nom!"

No IKEA. Learn or die.

And he did learn. Or I wouldn't be here.

Evolution. Survival of the fittest, through the ages. Into me. Into you.

Into Maya.

★ ★ ★

Wait a second. We're missing something.

Yes, the need to learn is baked into our DNA. Like graham in the crust.

Also, learning is hard.

Our brain doesn't like scary things. Or failure. This is how most people feel when facing new challenges.

Trepidation. Fear. Anxiety.

Do you see the paradox?

People don't *want* a challenge. They *need* a challenge.

★ ★ ★

At my peak, *Coach—freaking—Murphy*. My mastery was still only a 9 (of 10).

Sure, if I had stayed at U8 with the goofy boys chasing plastic bags as they blew across the fields? Yes, I had mastered that stage, 10 out of 10.

If the boys had stayed seven! They didn't.

They grew. The parents grew. The competition grew. Every year was different. U10. More serious. U12. More challenging in every way.

Before each new season, I felt the way Maya felt.

Trepidatious.

Followed later by the thrill of mastery. Over and over, in a never-ending cycle.

Chapter 19

The Boring Stuff

Do me a favor. Skip this chapter.

I'm serious. Skip to Chapter 20!

Why? Some details are boring. So, flip to the delicious picture. Drool a bit. Then, flip past.

Come back to this later, when you want even more.

★ ★ ★

Can we *manipulate* mastery and autonomy?

Manipulate is an icky word. It has that nasty aftertaste from the Velvet Tyrant.

If it's for the good of the individual, the team, the company, the customer?

Then it's leadership.

★ ★ ★

Option one. Decrease the challenge.

Do this when the task is too hard. Even with strong performers, sometimes, the work is too hard. People aren't great at everything. Snowboards and kiteboards.

They become afraid. They feel overwhelmed.

Engagement drops.

So, bring in other resources. Mentor. Coach. Break the task into parts. Shift the calendar. Whatever it takes.

This doesn't increase mastery. It just *feels* like it. Because the difficulty went down.

Engagement goes up.

★ ★ ★

Option two. Increase mastery.

This is my go-to. The long game with the highest rewards.

We find new challenges. Promote a growth mindset. Focus on the future. Praise new behaviors—especially growth.

Grit, plus growth mindset, is the wombo-combo of excellence.

Like my plan for Maya.

★ ★ ★

Option three. Lower autonomy.

Why would we intentionally lower autonomy?

Because Spock.

"The needs of the *business* outweigh the needs of the few, or the one."

Company restructure? Canceling a project? Organizational shifts?

Sometimes, autonomy isn't our biggest concern.

Do what you gotta do. Then fix it later.

★ ★ ★

There's a fourth option. We can't get there just yet.

First, some caveats.

Don't forget about low performers!

Engagement? Mastery, autonomy, purpose? That's for strong performers. The majority of your staff.

Engagement is a secondary concern for low performers. Unless it's the root cause of the drop in performance. That's a story from my other book.

However you do it, slay that elephant. The big, spectral, scary one.

And don't feel bad about it either. If Dumbo were as scary as this elephant, he wouldn't be the hero.

He'd be a villain! With a simpler name.

Fred.

★ ★ ★

The work is the work.

I like to hire superstars. Then I push them to become even stronger.

They are amazing, and amazing people want to work on amazing stuff.

Cutting-edge tech. Advanced algorithms. Particles and shaders. Beautiful landscapes. Character animation.

Because, DNA.

Millions of years of evolution. From Curtok SpearChucker, down to home-made graham crackers.

Golden brown sweetness with the perfect number of holes. A heavenly melt-in-your-mouth texture. Beautiful to look at. You almost don't want to eat them. Almost.

Not all work is awesomely interesting. Sometimes, it's spread-sheets, tax documents, and updates.

Even if the job is a dream come true. Even if you're building games or doing cutting-edge research.

Eventually, your staff will have desires you can't address.

Maximize their engagement anyway.

★ ★ ★

Sometimes, other leaders ask me.

"How long can you keep them engaged? Before they get bored."

My answer is always the same.

"As long as I can."

Chapter 20

Mastering Autonomy

I had a problem.

Yellow Team was too specialized.

It didn't start out that way. My original goal was to centralize the messy, hard engineering tasks. Pull the low-level tech into one place for efficiency.

Yellow loved it. The other teams loved it too.

Worked great for years. Until it became a problem.

Yellow was so good in some areas of the tech that other teams forgot how that stuff even worked. They became *too* reliant on Yellow.

That's when the problems started. Tech issues started backing up. Yellow went from helper to blocker.

I wondered.

What if I let developers swap to Yellow? Just temporarily. For three months at a go.

What if I made the whole thing voluntary? Took advantage of autonomy.

I gave engineers a choice.

"Who wants to join the Yellow Fellowship? New people. New projects. New tech. Lots of things to learn. Temporary, and voluntary."

They loved it.

It increased autonomy. Boosted mastery. Disseminated knowledge.

Win, win, win!

★ ★ ★

Option four is to raise autonomy.

Maximize autonomy with the small stuff.

Day-to-day work. Picking tasks. How they do the work. Who they collaborate with. The order of operations.

The details matter to the ones doing the work.

Like when I was a coach. I made lots of *little* decisions. How I talked to players. Which drills I ran. The way I engaged with them.

This gave me the feeling of autonomy.

Fus-Roh-Dah!

★ ★ ★

A word of caution.

Autonomy is not free. Not like praising behaviors or recaps.

It takes planning. Coordination. It creates disruptions. Moving projects or people. Collaborating across teams, across disciplines.

Especially for big stuff.

★ ★ ★

Remember my Team Shuffle? I asked that crazy question.

"Want to shuffle to a new team?"

That question raises autonomy. And it's a metric crap-ton of work.

Restructuring teams. Shifting projects. Physically moving desks!

Hundreds of conversations, each creating new opportunities for critical information to get lost.

Confusion!

Whisper loves the Team Shuffle. In fact, he loves any sort of autonomy.

Autonomy breeds confusion.

Yep. Some of this material is contradictory. You increase autonomy, and up goes confusion.

Too bad.

Leadership is hard. Unless you're okay with mediocre results.

If you want exceptional teams? You have to master all three pillars at the same time. Ruthless clarity. Strong performers. And relentless engagement.

All at once. In balance, like flow.

Chapter 21

The Flow Channel

Sometimes people get stuck.

They're working on a project they don't like. They're not vibing with their peers. Their skills are mismatched. They're bored.

That's a bad sign.

Engagement will drop. Performance will drop. Maybe they leave. Worse, maybe they stay. Becoming a quiet, low performer, draining the life from your team.

So, what do we do?

Wrong question again. The right question is:

Are you relentless about engagement?

Find your own crazy, bonkers initiatives. Let people pick tasks. Ask for volunteers. Set up a fellowship. Run a team shuffle.

That's the kind of stuff I do. And it matters. They tell me all the time.

"I feel heard."

More importantly, they show me!

★ ★ ★

You might not have the influence to pull off Team Shuffles or fellowships.

That's not the point. The point is this.

You have influence! Get serious about how you use that influence. Increase mastery. Raise autonomy.

Keep everything in balance. No matter how much effort it takes.

★ ★ ★

It's like the flow channel.

The first principle of game design. The day-one topic for my Master of Game Design students.

Flow has four requirements. The first three are easy to understand: immediate feedback, clear goals, and no distractions.

Then, the hard part—a perfectly balanced difficulty. Something that's "just right" for me.

Goldilocks OutOfTheBox.

Game designers spend their careers trying to keep players in the zone. And that gratuitous educational content also applies at work.

Mastery and autonomy require perfect balance, like flow.

★ ★ ★

Six months prior. I told Maya, "I have a big challenge for you. It's outside your comfort zone. What do you think? Do you want it?

I'd worked with Maya for years. We'd danced this dance before. She knew what I was asking.

She'd have a period of reduced mastery. Increased anxiety. She'd have to learn. Grow. There'd be months of challenging work.

It was a big ask.

Also, we knew how it would turn out.

The question wasn't *if* she could succeed. The question: was she up for this challenge? Did she *need* it?

The choice was hers.

Maya looked away. She thought for thirty seconds. Longer. Then she gave her answer. Resolved, with quiet conviction.

"Yes. I want it."

Six months later. She's sitting in my office. In the same exact chair. Telling me she felt "trepidatious."

Which meant. Her mastery had dropped. She was outside her comfort zone. She was kinda stressed.

It also meant. Her autonomy was *maxed*. Ten out of ten.

Maya was engaged. Committed to success. Climbing the steep trails of Mount Awesome.

That's why I was happy.

Chapter 22

Aligning The Vectors of Purpose

Relentless engagement is where great leaders shine brightest. Now, for the most important piece.

Purpose.

★ ★ ★

My boss scheduled a meeting with a weird subject.

"Planning Meeting"

It seemed random. No agenda, no prior discussion. Just this ominous thing looming in my calendar, "PLANNING MEETING!"

It sat there, taunting me. Then it got bumped to another day.

By the time we finally met, I was good and concerned. No, that's not right.

I was *trepidatious*.

The meeting started. She got right to the point.

"I have a proposal for you."

Aw, man! I hate it when they say that.

She said, "I want to make a big change."

Here it comes.

She said, "We are restructuring the studio. Changing leadership. And I need your help. I want you to pick up two additional departments. Quality and art."

Woah.

Whatever I was expecting from "Planning Meeting," this wasn't it.

I didn't know what to say.

★ ★ ★

Let's recap.

Mastery is measured against the task at hand. The skills you need versus the skills you have.

Autonomy is measured against the perception **of choice.** Autonomy is a question of belief.

Both can be measured with simple numbers. Sometimes high, sometimes low.

★ ★ ★

Purpose is different.

Take any random human.

There's so much going on in their lives. They are students. Lovers. Friends. Workers. Leaders. Parents.

They play games, watch videos, read books. They are chefs, artists, and coders.

They snowboard and kiteboard and surf.

Why do they do these things? Because they want to. Or have to. Or need to.

They *choose* to.

Desires, demands, or responsibilities. Add all that crap together. Across all the parts of their life. That's their purpose. The things that drive them to do what they do.

Purpose is motivation—that which drives action.

★ ★ ★

The question isn't how *much* purpose we have. It's a question of *alignment*.

How aligned is *your* purpose with *my* purpose?

It's about vectors.

Let's doodle.

★ ★ ★

Grab a crayon.

Take a piece of paper. Draw a line with your crayon. Any line will do.

Now, put a pointy hat at one end. Make it adorable, like the Hogwarts Sorting Hat.

Okay, forget the hat. Just draw an arrow.

This arrow has direction and length. We call it a vector.

And because it's on a piece of paper, we say it's two-dimensional.

Spice it up. Your crayon is now a magical wand. It can draw lines in the air!

Use it to draw a vector of Lionel Messi kicking the winning goal in a World Cup soccer match.

Do you know how much Messi is worth? He's the greatest soccer player of all time. The GOAT. Worth almost a *billion* dollars.

The GOAT has scored almost 900 goals in his career.

And people try to replicate those shots. Millions watch his videos. Hopeful.

They go outside. Place their ball on an empty field. They practice kicking. Again and again and again. Trying to mimic him.

Now you walk up.

You're watching Kicky Hopefulton. You take your magical crayon. You draw a vector in the air.

It starts at the ball and aims toward the net.

You say, "That's how Messi did it. Do it like that, Kicky!"

Hopefulton kicks the ball.

Then you take your magic crayon and draw another vector. Showing where theirs went.

Now you have two lines—two vectors. They're floating in the air. You can walk around them. Touch them.

You gotta touch them, right? It's a pair of magical lines floating in the air! Who wouldn't touch that?

And more important than touching them, just look at them.

You're looking at these two vectors. You can see how close or far apart they are.

How **aligned** they are.

This closeness between Messi's shot and the hopeful contender? That's how we measure purpose.

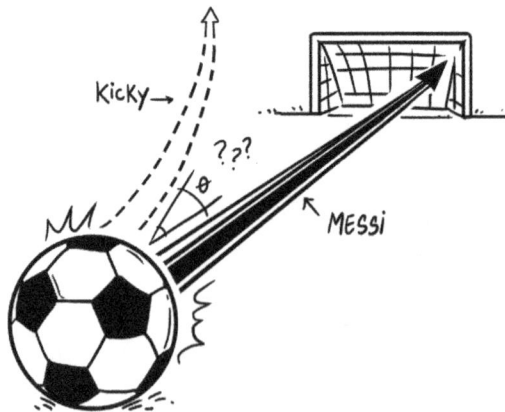

Every employee has their own motivations.

A sense of duty. Providing for their family. The love of a challenge. A creative outlet. The pursuit of excellence.

Maybe they want to change the world.

Or maybe none of that. We can never know their reason for getting up in the morning.

We don't need to know.

All we need to know is whether their purpose is aligned with our purpose for the work.

★ ★ ★

With these two vectors. Our goal as leaders is easy to understand.

You want to increase engagement?

Simple.

Align the vectors of purpose.

Chapter 23

Mi Norte, Su Norte

My boss had just dropped a bomb.

"I want you to pick up two additional departments."

She walked me through it—starting with the core problem. The recent departures. The headwinds that could destroy us.

Then, she shared her vision.

A new organization. New roadmap. Different reporting structure. How things would work. My role in it all. Everything.

She prepared slides and a compelling pitch.

She said, "You don't have to do this. Only if you want it."

"It's my choice?"

"It's *your* choice."

★ ★ ★

My boss gave me autonomy—exactly like I did with Maya. Kudos and well played, madam!

Challenges. Mastery. Autonomy.

Blah, blah, blah.

That's so five minutes ago.

★ ★ ★

My boss painted a picture. Conveyed the worries. The challenges. The conviction. The opportunities. The vision.

Know what we call that?

If you guessed "purpose," then have a homemade graham cracker.

My boss shared her purpose. Her vector. She made her north my north.

★ ★ ★

Do you know why nerds love vectors? No, it's not because it looks like a phallic symbol, not at all like that Amazon logo.

How it swoops, curves, with the rounded end. Never mind, I'm sure that's just me.

Nerds like vectors because they can have many dimensions. Not just two or three. They can have hundreds, thousands of dimensions.

It's hard to think about a vector with 1,000 dimensions, and mathematically, it's dirt simple. And very practical.

Vectors are literally going to change the world. Because, AI. You know, large language models. ChatGPT, Gemini, Claude.

AI is made with a simple recipe.

Fill a large bowl with vectors. Knead them until they have ten thousand dimensions. Sprinkle in some simple math, to taste.

Now toss that nerd-loaf in the oven at 450 degrees. Bake until golden brown, probably about four months. Sprinkle a sugary interface on top.

Voilà!

Nerd-loaf can tell you the capitals of all fifty states. Help your teenager with their "new math." And diagnose that weird fleshy growth that you definitely did not take a picture of.

All because of vectors. Running on a silicon wafer.

Which is mostly just sand.

★ ★ ★

My boss made a lot of effort. And took a big risk, giving me the power to say no. To cancel all of those plans.

The aim wasn't to "convince me." Or get me to *understand* what was being said. It was more than that.

Much more.

She wanted me to internalize the vision. The good, the bad, the ugly. Every part of it.

She wanted me to grok it. Deeply and truly.

More.

She wanted *her* purpose to become *my* purpose.

Mi norte, su norte.

Chapter 24

They Don't Know

Purpose is motivation. The reasons we act. And that can't be mapped to a simple number between zero and ten.

It's a vector. Like the doodle with our crayon. The magic dreams of Kicky Kickpants.

To increase engagement, align the vectors.

Ours, the company's, the employees'.

★ ★ ★

Now, let's talk about mission statements.

You know, those hyped-up, buzz phrases you see on coffee mugs. On shirts. And pens with not enough ink.

Most companies have mission statements.

- Google: "organize the world's information."
- Amazon: "to be Earth's most customer-centric company."
- OpenAI: "ensure that artificial general intelligence (AGI) benefits all of humanity."

I see. AI is for *our* "benefit."

"For all of humanity."

Right. Especially the AGI part. Where computers become smarter than humans. At everything.

Nothing to worry about.

(*sips coffee*)

Um. Mission statements.

They're nice, I guess. For building brands.

The question is: do mission statements affect engagement?

★ ★ ★

Ask yourself.

Why did my boss go to all that trouble? Why the big pitch?

If mission statements were so grand, couldn't she have saved herself a lot of trouble?

Told me, "You gotta do it to meet the mission statement."

"It's for the *mission statement*? Count me in."

BZZZT!

Mission statements have a place. And that place doesn't include aligning purpose with the mission.

Not in the short term. Not today.

My boss knew that. I knew it. You know it.

Do *they* know?

★ ★ ★

Do your people know what the mission is?

No. Not "The Mission"—with a capital em.

I mean.

Do they fully and completely internalize the key aspects of what's needed in the business right now, such that their purpose is aligned with you, other stakeholders, and the company?

Was that the longest sentence in the book? In a game, we'd get an achievement.

You've unlocked: *Lost the Plot (and Kept Going)!*

Here's a less loquacious version.

Are their vectors fully aligned with yours? I suspect you know the answer.

And Whisper knows it too. In fact, the Master of Confusion is peeing his pants right now. We're about to expose his big secret.

Your people do not know what the mission is.

That's right.

Your strong performers. The ones doing the work. Building the products. Making the money.

They do *not* know what the mission is.

Not really.

This is another one of those elephants. A reality hiding in plain sight.

We got work to do.

Chapter 25

Which Why?

Remember BigMegaCorp? It was back in Chapter 1.

The team was *almost* exceptional.

We missed that little clause about needing the build ahead of time. Three weeks? Just kidding. Three days.

Let's see how that story began. Before we caught the mistake. Before we even started the work.

★ ★ ★

"Welcome, everyone. Thanks for coming."

So many eyeballs looking at me. All fearing the worst.

And why not? We had just met as a big group last week. Then, out of the blue, a *mandatory*, last-minute meeting? With the whole group?

It had to be bad. Layoffs? Maybe we got bought. Or something worse!

Seeing their fears, we softened things.

"Don't worry. It's not a bad thing."

Says you.

"Seriously. This is a good thing. It's just urgent. BigMegaCorp wants the impossible. They want three games ported, integrated, and ready to launch in ten weeks."

So, not a layoff?

"No. No layoffs. It's an unexpected gift."

Is that good or bad?

They were still playing catch-up. Trying to shift from fear of layoffs to ... whatever this meeting was about.

★ ★ ★

"BigMegaCorp has a new platform. They want our games on it. And they want them fast. We'll get to that. First, let's talk about the elephant."

Elephants are scary. Fred, not Dumbo.

"BigMegaCorp requested us. Moved us to the front of the line. We'll see millions of new customers. They'll play our games for years. Maybe decades."

Then a dramatic pause. Building tension.

"This is maybe one of the biggest opportunities some of us will ever see. Maybe in our whole careers."

Another pause.

"*If*. And this is a big if. *If* we can pull it off. Cause I'm not sure. It's a tight deadline. There are so many unknowns. We don't even have a contract yet. We might not succeed. And if we don't, they'll launch without us."

★ ★ ★

Back in 2009.

A leader was giving a speech. He held a microphone in one hand. A marker in the other. One of those big, fat markers. Like in kindergarten.

This leader was doodling on a pad of paper. The huge kind on an easel. He drew three circles. On each, he wrote one word.

On the outermost circle: "What."

On the second circle: "How."

In the center circle: "Why."

The speaker was Simon Sinek. And his talk was titled, "How Great Leaders Inspire Action."

Or as most of us think of it.

The Golden Circle.

I say "most" because you've probably heard of his talk. You've probably seen it. You might even have guessed where I was going.

Those three circles are iconic. It's still one of the most-watched TED Talks of all time.

Simon Sinek has that claim to fame because he made a very, very, very good point.

You want to align purpose? Get those vectors pointed in the same direction?

Start with Why.

Write that on your hand. Or your forehead. Or both, like Jim Carrey in *Liar Liar*.

Etch it with a laser on the back of your phone. Print it with a 3D printer. Set it on your desk. A visual reminder so you never forget.

Always start with why.

★ ★ ★

You probably knew that. Maybe, like me, it's already etched in your soul.

Great. Now let's take it further with a hard question.

Which "why" matters most?

★ ★ ★

Go back to the BigMegaCorp crisis.

Check out the ice-cold bucket of concepts I dumped on their heads.

Initial fears of layoffs. The new contract. The opportunity. The millions of players. The timeline. The possible impact on revenue. Years. A decade!

There were lots of possible whys.

Which "why" was *most* important?

★ ★ ★

Start with first principles—the fundamentals.

A leader uses influence to achieve results while maintaining trust.

So, you're me.

An awesome opportunity just drops in your lap.

What's the first thing you should do as a leader? I'm talking about before anything else. Before you plan, react, or jump into action.

The first step?

Figure out what results you want to achieve.

In this story, the millions of customers mattered. It also mattered that it was BigMegaCorp. And that they wanted three games.

Blah, blah, blah.

No!

What was most important?

The *date*! If we missed the date, BigMegaCorp would move on without us.

Everything else paled in comparison.

★ ★ ★

Start at the end.

Imagine you're on the finish line, grinning, clinking your glass of champagne. Then, work backwards.

What led to success?

That's the first leadership action. Be Goalstradamus.

It's the answer to most leadership problems. At once easy to understand. And in practice, harder than it seems.

I know this because I've coached lots of leaders.

They bring their crazy situations. Lay out the details. Explain their conundrum.

I ask, "What else?"

Again and again until they've told me everything. Then they look at me expectantly. Twins again, waiting for a solution.

And I can't help them yet.

"I understand the situation. And you've forgotten something."

"I did?"

"What *results* do you want? When this is all said and done. What are you trying to achieve?"

"Oh ... Good question."

Your first step is to figure out which result you want to achieve. Then work that backwards into a tangible, meaningful "why."

Whylock Holmes.

Chapter 26

Of the Mission, By the Mission, For the Mission

Let's recap.

We're working on relentless engagement. Commitment to success, which requires mastery, autonomy, and purpose.

Purpose is motivation—the reason for action. A vector with many dimensions.

A leader's job is to align those vectors. Get everyone's purpose pointing toward the same north.

To do that? Start at the finish line. Figure out what you're trying to achieve—which "why" matters most.

All so you can align their north with your north.

★ ★ ★

What if you can't be honest about the why? What if the results you're trying to achieve are sort of sneaky? Or protected?

It happens.

Maybe you're preparing for a restructure. Or being sold. Or something else that can't be talked about openly.

Sometimes you might not be able to tell the *whole* truth.

If you can't be authentic about the results you want? Or about the real why?

Then you're forked.

Think about it.

Aligning purpose isn't an intellectual exercise. We're not doing this for funsies.

We align it because it increases commitment. And commitment doesn't just mean energy. Excitement. Working harder.

No.

We want commitment to success (of the mission).

See that new part, in parentheses: "(of the mission)"? That's been implied all along.

And we never called that out. Because it's been assumed. It's been obvious.

Like if we ask AI to get rid of wars. We don't expect the answer to be, "eliminate all humans."

No humans, no wars. Problem solved from the AI's perspective.

BZZZT.

"Get rid of wars (while not harming humanity)."

The part in parentheses matters!

★ ★ ★

Engagement sharpens focus. Yields faster decisions. Simplifies collaboration. Nails the details. Increases creativity.

It's not just about energy. Or intensity.

It's about catching the mistakes before you lose $50M. It's about building the right product. Accomplishing miracles. Moving mountains!

So, if you're not being transparent about the part in parentheses? If you can't be authentic about the mission?

Then you're hitching a slow ride to Failtown. You're focused on output, instead of outcomes. You're gonna end up in Death Valley.

Without authenticity, you can't have exceptional teams.

Don't be mad at me—Mr. Messenger.

★ ★ ★

They sat there. All those eyeballs, now wide open.

"There are a lot more details. Which games? Which teams? Next steps? The impact to the roadmap? We will get to all that."

Then the pause.

Pauses break people's stray thoughts. Bring everyone back to center.

"The important thing is the date. And dates aren't our strength. Our strength is quality. Which is still important, as long as we hit the date. The date matters *most*."

Lots of stuff happened in that kickoff meeting. We discussed all the details. There were lots of questions.

And did we have answers to all possible questions? No, we did not. There were many unknowns about "what." And "how."

Those other two parts of the Golden Circle were still a bit fuzzy.

The center circle was spot on. We knew the results we needed to achieve. The why.

We had to achieve the date!

★ ★ ★

Hitting the date for BigMegaCorp had nothing to do with mission statements. It also wasn't about some aspirational yearly goal. Or a spreadsheet of items in a quarterly review.

Success is rarely about those things.

Success is about the mission in front of the team. The part in parenthesis.

If you want excellence. Exceptional teams. Then figure out which results matter most.

Decide the best "why" for today. Tomorrow. The near term.

For BigMegaCorp, the why was the date. Hit it or they'd move on without us.

That led to laser focus. Which was the only reason we had a shot. Weeks later, when we realized our mistake.

Three weeks left? Hah! You wish.

We had three days!

Chapter 27

Stories For Dummies

D o you want to affect others?

It's a serious question.

Do you really want to influence other people? Change their thinking? Impact their goals?

How about shifting their motivations? Affecting the reasons they act.

It's a weird way to think about it. So, let's say the quiet part out loud.

Affecting others is your job. So, get good!

Level up so you can unlock new skills, like the enchanted, double-barrel, Big-Freaking-Gun (BFG).

The Bazooka of Engagement!

★ ★ ★

Everyone was looking at me. Hundreds of people in a hot auditorium.

The heat from the spotlights burned my skin. My heart raced. And my mind blanked.

It was the fourth slide of my presentation, and I stood there blank-faced. With no idea what I was going to say.

Worse, it was one of those weird presentations where the slides kept advancing. Every 15 seconds.

Flip. 15 seconds. Flip … Flip … Flip.

It didn't matter that I was stuck. The unstoppable wheels of time kept turning.

Tick. Tick. Tick!

At that moment, I was kicking myself for volunteering. Auto-advancing slides? What a stupid idea!

Focus, Curtiss!

The auditorium was quiet. All eyes on Sir Chokes-a-Lot, with no words.

Worst of all.

It was the first and only time in my career that my mother was in the audience.

★ ★ ★

"Crap!"

I said later.

After I went home. Feeling humiliated. Realizing I wasn't as good as I thought I was.

And there was no time for self-pity. I had another speech the next day. Two talks in one conference?

"Not very bright. This one."

So I did what anyone would do.

I practiced even harder for tomorrow. It was material I knew better. There were no auto-advancing slides.

Just me, the theories of game design, and work that I knew like the back of my hand.

★ ★ ★

Tomorrow came. I delivered a flawless presentation. Then I gave the classic line.

"Any questions?"

That was before I knew the better question. "What thoughts, questions, concerns?"—that wisdom came later.

Fortunately, people asked questions anyway.

Which I handled like a pro, if I say so myself. Then, up came a hand from someone I'd only ever seen from a distance.

"Yes. Captain Barret?"

In his dress uniform. Polished. Professional. Perfect.

"Curtiss. I've seen you talk before."

He knew who I was? Nerd-bumps!

He continued, "You talk about game design. And I'm wondering. How come you never talk about *stories*?"

Stories? Wait. What?

I was talking about *flow*. The psychology of engagement. What makes great games great. I hadn't said anything about stories.

Oh! That was his point.

How come I never talk about stories?

"Um ..."

Call incoming, from Professor Holdplease.

Ahah!

"Well, Captain."

No first names for me. Best behavior only!

"Well, Captain. I'm not sure stories are required for games. Tetris doesn't have a story. Right? And even for games with stories? Many players skip them. That suggests stories may not be critical for engagement."

Nice recovery!

He nodded and sat down. I breathed a sigh of relief. Then later, when I went home.

"Crap!"

★ ★ ★

Twice in two days. One, a disaster. The other, a near miss. Both in public.

I was crushed. And also, the captain had given me an idea. A way to channel my defeat. It was time for me to learn about stories.

Ever try searching the internet for the word "story?"

Yeah. Not helpful.

Best you'll get is a bunch of stuff I didn't understand. Arcs. Storylines. Character development.

Blah, blah, blah.

I needed *Stories for Dummies—The Nerdy Guide for Double-Shanking Newbs*!

Eventually, I found *Tell to Win* by Peter Guber. CEO of both Mandalay Entertainment and Sony Pictures.

There, in the early chapters, was the answer I needed. A recipe that was dumbed down to my level.

1) Question/ Challenge
2) Emotional Journey
3) Galvanizing Conclusion (preferably with a twist)

★ ★ ★

It was a simple recipe. Easier than the one for AI or graham crackers.

Perfect for an engineering mind. All I had to do was add the ingredients in order.

So, that's what I did. Obsessively.

Emails. Texts. Articles. I used the story formula everywhere. When writing. Talking with my team. Updating my boss.

Even with my wife.

"Stop doing that," she said.

"Do I have to? I'm earning my 10,000 hours. One day, I'll have my own special. Mr. Rogers' Nerdy-Neighborhood!"

It made her laugh, despite her protestations. The yin to my yang.

★ ★ ★

I practiced. And practiced. And practiced.

Months passed.

Until the next conference. The biggest presentation of my life. Ninety minutes of "Why Games Work—the Science of Learning."

It was the largest auditorium. I was the featured speaker.

Well, sort of. I mean, I had the coveted first slot. 8 AM, Monday morning. It was pretty scary.

Clever me, I brought donuts. Five dozen donuts, for 500 people! What a dork.

A nervous dork. Jittery, like the Navy SEAL.

Every seat was full. People lined the walls. Peering in from outside the doors.

It was my time to shine. And shine I did.

I had stories now. Lots of stories woven into my slides. In, out, I was zooming.

Then. The mic went dead.

"Hello? Is this on?"

Tap. Tap!

Nothing. I began shouting.

"CAN YOU HEAR ME IN THE BACK? LET'S KEEP GOING!"

I didn't even lose my place. The stories kept me on track. It was easy. Almost effortless. I was in flow.

At some point, they gave me a new mic.

Tap. Tap!

"IS THIS ON? Oh. That's loud. Okay."

Picked right back up. Crushing it. Then, the second mic went out. I kid you not.

"Wait ... is this dead too?"

Forget the mic.

"WE'RE GOING! CAN YOU HEAR ME? OKAY."

Grit. Perseverance. I was galloping towards the final lap as someone tried to hand me a third mic.

"ARE YOU KIDDING? I DON'T THINK SO. ANALOG BABY!"

And I finished strong. A galvanizing conclusion, with a twist.

"Do you see it? Good games work for the same reasons that good learning works. No more broccoli ice cream for you!"

I dropped the mic I didn't have.

Claps, applause, and later that night, during the speakers' dinner, I was lost in thought. My heart, still racing, ten hours later.

I heard my name.

"And the winner for Best Tutorial of the 2011 conference? ... Curtiss Murphy!"

They took my picture and handed me a crystal trophy. It was heavy.

This is the power of stories.

From BlankFace Jim. To Crystal Avenger.

One. Start with a question or a challenge. Like I did earlier. Forgetting my place, with my mother in the audience.

Two. Share the emotional journey—Capt. Barret called me out. Then hours of practice, like Rocky, shadow boxing his way up those famous stairs. The mics went dead. Twice!

Three. Finish strong, with a galvanizing conclusion (with a twist). I didn't just finish my presentation, I took home gold!

That was a story. Using the recipe. With a twist.

Stories are one of the most powerful tools you have as a leader. They move people. They convey purpose better than almost anything.

Practice telling stories in everything you do. Even if the moment isn't important. Keep practicing.

For when it matters!

When it really, really matters—you'll be ready. Able to tell a great story that connects directly to the mission.

Tell stories that align the vectors of purpose. For your strong performers.

That's the Bazooka of Engagement.

Now, Bernoulli.

Chapter 28

The Pull Effect

I hope you've found some gems in all this material. Diamonds that have made you a better leader.

I hope you're already applying them in your mind's eye.

As we head into the home stretch, I want to share one final technique.

This one's advanced.

It's extra. Like a bonus fry at the bottom of the bag. The unexpected joy of realizing there was one left after I thought I'd eaten my very last peanut butter cookie.

You can skip it.

The chapter, not the cookie. Obviously.

★ ★ ★

Or, we can keep going.

Okay.

Bernoulli's Law of Leadership. Or said simpler.

The Pull Effect.

★ ★ ★

Imagine with me.

It's early evening. It's been a warm day. And the upstairs of your two-story home is plenty hot.

Your air conditioner's on the fritz. No AC.

Finally, the sun starts to set. The outside temperature begins to drop.

You're standing there. In the oppressive stillness of the master bedroom. Down at your feet is your lonely box fan.

Now, the question.

Where do you place your fan?

What's the best way to cool the master bedroom?

★ ★ ★

In 1738, a Swiss mathematician made a strange discovery.

He took water. Just regular 'ole water. Then pushed it through a tube. Faster and faster until he noticed something odd.

The pressure went down.

Not the force of the water slamming into the rest of the water. That's different.

We're talking about the side pressure. The outward push of the water on the walls of the pipe.

When the water moved faster, the side pressure dropped.

That's not intuitive, it's mind-blowing. You'd think faster means stronger. More water pressure. More force against the pipes.

And that's not what happened.

Daniel Bernoulli discovered that the water wasn't using its energy to push outward. It was using its energy to move forward!

The faster the flow in the pipe, the less it pressed sideways.

That's Bernoulli's Principle.

And weirdly, the same thing happens with air.

When air speeds up, the pressure drops. It creates an area of low pressure behind.

Chalk one up for science.

★ ★ ★

So, who cares?

Well, the Wright Brothers cared. Bernoulli is responsible for air travel. And perfume bottles. And carburetors.

And also, a cool thing you can do with that box fan in your upstairs bedroom.

★ ★ ★

Bernoulli's Principle holds the key to our earlier riddle.

Where do you place your fan?

You've probably heard the answer before. You'd naturally put it in the window to pull in the cool evening air.

Unless you have small windows. Too small for that fan.

Then, put it in the doorway. Face it outward, with the air blowing toward the kids' rooms.

Or in my case, the guest room. Because kids grow up. Move out. Travel 3,000 miles away. And call you on Father's Day.

I miss the kids being down the hall.

★ ★ ★

Speaking of kids, here's some gratuitous educational content.

Ever notice that your house is often hotter on one side? It's usually the side with the kids' rooms. Because builders try to put the parents' room on the cool side.

No kidding.

And it's easy to guess which side will feel hottest. It's the side with the morning and afternoon sun. The sun warms up the walls, the rooms.

Then the kids complain, because their room's always hot.

So one side *is* cooler. Usually, the side with the master bedroom.

Open the windows. Put the fan in your doorway, pointing toward the kids' rooms. And if you have two fans? Put one in the kids' doorway too.

From cool to hot. Physics! And this isn't a physics class.

It's a metaphor.

★ ★ ★

So, now the interesting question. Why?

Why *inside* the doorway?

I mean, it's awfully inconvenient. Making it hard to walk and such. So, let's break it down.

The fan pushes air forward. That part's obvious.

Then comes the magic of Bernoulli. As the fan starts pushing the air. Faster and faster. It creates an area of low pressure behind the fan, inside the master bedroom. That creates a sucking behavior.

Exactly like a vacuum cleaner.

Or a thunderstorm, rolling across verdant green hills. Unless you live in Southern California. Where it never rains.

"Why did I move to the desert?!"

"But technically, it's semi-arid."

"Zip it, WeatherVane! It hasn't rained in six months."

I would never call someone a WeatherVane. That's rude, even if they were a But-Head.

And I truly did miss the weather when I first arrived.

Real thunderstorms, you know? Where we describe rain with words like "deluge." Where we worry about the shingles blowing off the roof.

Trees crashing through the attic.

That's when you meet your real neighbors. The ones with the chainsaw. The ones who are super helpful after a hurricane.

And if you're wondering.

Why all the goofy tangents in this chapter? Because ... "A spoonful of sugar helps the medicine go down." —Mary Poppins

★ ★ ★

So.

Your box fan.

It pushes a little bit of air through the blades. The air speeds up. Which creates a vacuum behind it.

That vacuum *pulls* a bunch of additional air. With a big sucking sensation.

Whoosh!

The faster the air, the stronger the pull.

The Little Fan That Could? He absolutely can!

However little air your fan is pushing forward. That vacuum is putting in serious work. Creating a low-pressure area that sucks cool, fresh air through the window.

Tada!

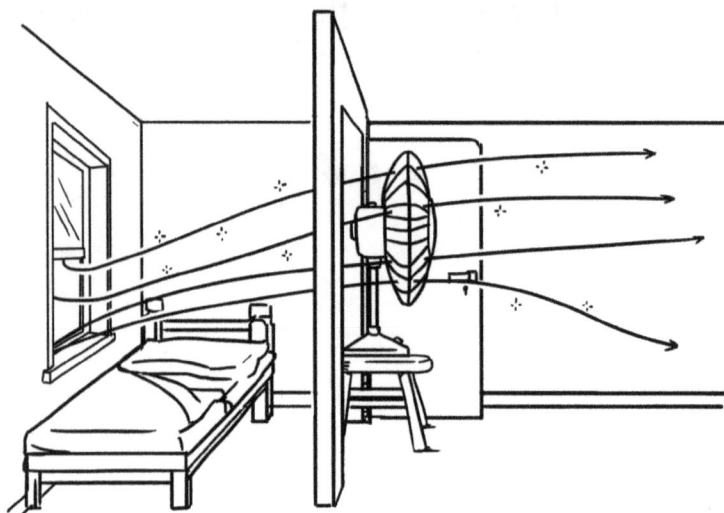

Now. What does that mean for us?

Keep the visual in mind. The bedroom window is open. The fan's in the doorway. Pointing down the hallway.

Think about how smooth that air is flowing. How fast things are cooling down. The best that physics has to offer!

Until the HVAC dude shows up.

★ ★ ★

So your fan is doing its job. Then you decide it's taking too long.

You move the fan out of the doorway. Set it on your dresser. Or a stool.

You put the fan in the middle of the room. Facing toward your bed, because you're hot!

What do you think happens to the air?

The same physics apply. The fan pushes some air. The air moves faster. This creates low pressure behind the fan. Which creates a vacuum. Which sucks air in.

So what's the result of just pointing the fan toward the bed? A whole lotta nothing, that's what.

It's not sucking cold air in from the window. It's not blowing hot air out to the hallway. The air just moves around the room, in a giant circle.

So, yes, it's blowing on you. Which does cool you down a bit. And mostly, it's just a big, agitated vortex, blowing nowhere.

Hold these two visuals in your mind while you read this next sentence.

You are the fan; the room is your team.

★ ★ ★

Do you see it?

You are a fan, blowing air in the room that is your team.

Your actions. The way you use your influence. It's like pushing air. Faster and faster. Creating low pressure behind you.

And if you're sitting on the stool? Facing in toward the bed? Then you're just swirling air in a hot, stuffy room.

The Vortex to Nowhere.

That's what most leaders do.

They use their influence to push air around a room. Bloviating. Or maybe, *Blow*viating!

Don't be Chief Blowhard.

★ ★ ★

What does this mean practically speaking?

It means thinking ahead. Make sure you're not just pushing air around a room.

Swirling your team. Fighting over small things, moving no one, nowhere.

Put yourself inside a doorway, pointing down the hall!

Work toward a goal that isn't at odds with what's already happening. Somewhere with less resistance.

Then, start leading in that direction.

You sit in the doorway, pushing air into the hallway. As it speeds up, you get the vacuum, which pulls air towards you.

The Pull Effect.

★ ★ ★

Forget the metaphor. Let's make this practical.

Remember the Team Shuffle?

I saw a problem that needed solving. A drop in engagement across all departments, due to a recent restructure.

I started slow, months ahead of my target date.

I put myself *in* the doorway.

Starting at the lowest speed—a gentle breeze. Pointing the direction I needed us to go.

My early steps involved strategizing with my boss. I shared my concerns about the decrease in autonomy. The drop in engagement.

Slowly. Carefully. I turned up the speed of the fan.

I built alignment on the problem with engagement, autonomy, and interpersonal conflicts. I shared more aspects of my plan. With my boss. With other senior partners.

The vacuum began pulling energy in the right direction. Then I offered a suggestion.

"It's been a while since our last Team Shuffle. Might be time for another. What do you think?"

★ ★ ★

An additional notch on the fan. Faster.

I included other stakeholders. Gathered their thoughts. Explored options.

The air moved faster. Created an area of low pressure behind me, pulling more people forward.

Weeks passed.

Another notch. Faster.

I expanded the circle. Engaged with mid-tier leaders, one by one. Partnered across departments.

A larger vacuum. A stronger pull until something strange happened.

People began adding their own ideas into the mix. Details about the shuffle I hadn't considered. They solved problems without me even knowing.

Finally, my big meeting. The question for every person in my group.

"Want to shuffle to a new team?"

The fan's on max speed.

The air's fast. It's pushing some people forward while also creating an area of low pressure behind. Moving people in the right direction.

After that?

Almost everyone was aligned. People figured out seating. Solved roadmaps. Coordinated move days.

It had momentum of its own. I was practically a passenger at that point.

The target date arrived. And ironically, the whole thing happened in a flash, done before lunch.

The mountain moved.

★ ★ ★

No one had pre-existing attachments to a Team Shuffle.

"Curtiss is weird. That's kind of his thing."

Pointing outside the room led to less resistance. No entrenched concerns to overcome. Just a smooth flow of energy, escalating in the right direction.

Without flailing, yelling, or arguing. No fights over territory. Almost no resistance at all.

From overwhelming to effortless.

Just a fan in a doorway. Pointing down the hall on the advice of a nerd from the eighteenth century.

★ ★ ★

Pick a direction outside the room. Push a little air. Create an area of low pressure.

Then let the physics work for you.

Once the vacuum begins pulling people in the right direction. Increase the speed. Rinse. Repeat.

The Pull Effect.

For leaders who want to move mountains, without making waves.

Now, the ending of our story.

Chapter 29

Three Days To Impossible

Back to that meeting. The BigMegaCorp crisis.

"We have a problem."

I explained the hidden fine print. We didn't have three weeks. We had until Friday.

And it was Tuesday.

High stakes. Tens of millions on the line. Three days to go.

So many leaders, learning that months of effort were flushing down the drain.

I said, "What's the options? What date could we hit?"

Then I opened things up. The teams talked. Fluid and smooth.

No one raised a voice. No one pointed a finger. Everyone stayed calm, looking for solutions.

They asked about the contract. The *exact* language—what was due when.

Each team gave an update. Then, Clarence said something surprising.

"I don't think we should."

I said, "We shouldn't what?"

"We shouldn't ask for an extension."

"Tell me more."

"My team was freaking out about that date. We stayed *way* ahead of schedule. At this point, we're almost done. My team can make it."

"In three days?"

"I think so."

"What about everyone else?"

One by one. The leaders sounded off.

"Yes."
"We're close."
"We can do it!"

Each team agreed. Each discipline, too. Art, engineering, testing. Production. Design. More.

They agreed and went to work.

Three days later, we delivered the builds to BigMegaCorp. Weeks ahead of the original plan. With only days of notice.

They did it!

That was years ago.

Those games are still live. Played by millions. With incredible revenue! It was a plot twist that changed the trajectory of the company.

The twist wasn't a bunch of overtime. In fact, I don't remember that anyone worked an extra hour. They didn't need to, they were ahead of schedule from day one.

Three weeks early on a ten-week job!

Highly engaged, strong performers who collaborate well. Focused on the mission. With ruthless clarity. Committed to success.

Doing the impossible!

This is the power of **unstoppable teams**.

★ ★ ★

Speaking of adorable ... Here's my two mutts. Master Candy, The Wise. And Young Dottie, Launchpad Leaper.

They did their part too. Licking, begging, and reminding me that playtime is important too!

Is there more? Of course.

Outcomes over output. Being vulnerable. Credibility. Interviewing for excellence. Improv for leaders. Thriving amidst crisis. The perils of being a But-Head. Peach eats the apple.

There's no end to the leadership journey.

And for now? You have enough to work on.

- Ruthless clarity
- Strong performers
- Relentless engagement

These things took me a long, long time. I'm hoping it won't take you that long.

I'm hoping I've sped you up. Pulled you in the right direction. Gently, effectively. Like Bernoulli.

The rest is up to you.

Kill confusion.
Drive engagement.
Build an unstoppable team.

And one more thing.

The End

How many people leave reviews?

Only one in a hundred. They don't think it matters. Which you know is wrong.

Strong performers need feedback. Including me. So, please leave a review. Click those stars.

And give a shout to Ivano for the lovely illustrations. Hand-drawn with care.

Praise the behaviors you want repeated.

Fus-Roh-Dah!

www.ingramcontent.com/pod-product-compliance
Lightning Source LLC
Chambersburg PA
CBHW031926190326
41519CB00007B/430